D1164684

TIME
OF OUR
LIVES

The Science
of Human Aging

TOM KIRKWOOD

OXFORD
UNIVERSITY PRESS

OXFORD
UNIVERSITY PRESS

Oxford   New York
Athens   Auckland   Bangkok   Bogotá   Buenos Aires   Calcutta
Cape Town   Chennai   Dar es Salaam   Delhi   Florence   Hong Kong · Istanbul
Karachi   Kuala Lumpur   Madrid   Melbourne   Mexico City   Mumbai
Nairobi   Paris   São Paulo   Shanghai   Singapore   Taipei   Tokyo   Toronto   Warsaw

and associated companies in

Berlin   Ibadan

Copyright © 1999 by Tom Kirkwood

First published by Oxford University Press, Inc., 1999

First issued as an Oxford University Press paperback, 2000
198 Madison Avenue, New York, New York 10016

Oxford is a registered trademark of Oxford University Press

Library of Congress Cataloging-in-Publication Data
Kirkwood, Tom.
Time of our lives: the science of human aging / by Tom Kirkwood.
p. cm.
Includes bibliographical references and index.
ISBN 0-19-512824-9 (Cloth)
ISBN 0-19-513926-7 (Pbk.)
1. Aging—popular works. I. Title.
QP86.K52   1999
612.6'7—dc21   98-46932

1 3 5 7 9 10 8 6 4 2
Printed in the United States of America
on acid-free paper

*For my mother, Deborah Kirkwood, and
in memory of my father, Kenneth Kirkwood*

Dangerous pavements.
But I face the ice this year
With my father's stick.

Seamus Heaney, '1.1.87'

CONTENTS

Preface

Today's older people are the vanguard of an extraordinary revolution in longevity that is radically changing the structure of society and altering our perceptions of life and death. Improvements in sanitation, housing, healthcare and education have resulted in huge increases in expectation of life. Countless lives are now lived to the full that might otherwise have been cut short. The price for this success – and make no mistake, it *is* a success – is that we now face the challenge of ageing.

Time of Our Lives is about the science of human ageing – one of the last great mysteries of the living world. Questions like 'Why do we age?', 'How does ageing happen?', 'Why do some species live longer than others?', 'Why do women live longer than men?', 'Do some parts of the body wear out sooner than others?' and 'Why do women have a menopause halfway through their life span?' beg for answers. The more practical question 'Can science slow my ageing process, or help me age better?' concerns us all.

For all of these reasons, human ageing is at the forefront of scientific, medical and social research and of political thinking as never before. If we are to meet and overcome the immense challenges of adjusting to the worldwide demographic revolution, with all that this means in terms of longevity, economics and, above all, quality of life, we need to be armed with better knowledge.

Time of Our Lives is written to be intelligible to a reader who has no training in science, but an interest to know. It is also written for those whose daily work brings them increasingly into contact with older people. It even, I hope, has messages for the policy-makers, those we elect to lead us into the uncharted territories of a greying world.

It is not a textbook and yet it goes to the heart of current research. Although I have been at pains to make every part of the book informative and intelligible to the lay reader, I have avoided oversimplication so that the real issues can be understood. I make no excuse for asking the reader to work a little harder in some chapters than others to follow material that may be unfamiliar. My side of the bargain is that I have also worked hard to make the science as accessible as possible, using everyday examples to help. The focus is on ageing, of course, but quite a lot of other interesting science has been included to set the stage. In writing for a diverse readership, there is an unavoidable danger that one will sometimes explain a point at greater length, or at a simpler level, than the reader requires. Where I have failed to get the balance right, I hope that this will not distract too much.

Many colleagues have helped my research over the years. Particular thanks are due to Steven Austad, Thomas Cremer, Ioan Davies, John Grimley Evans, Caleb Finch, Claudio Franceschi, Leonard Hayflick, Robin Holliday, Tom Johnson, Axel Kowald, Gordon Lithgow, George M. Martin, Ed Masoro, John Maynard Smith, Leslie Orgel, Linda Partridge, Olivia Pereira Smith, Chris Potten, Patrick Rabbitt, Francois Schächter, Daryl Shanley, Jim Smith, Richard Sprott, Raymond Tallis, Roger Thatcher, James Vaupel and Rudi Westendorp. I would also like to express my appreciation to the organisations that have supported my research at various times, particularly the UK Medical Research Council, Research into Ageing, the Wellcome Trust, the Dunhill Medical Trust and the UK National Biological Standards Board.

My agent, Felicity Bryan, first suggested that I write this book and has nudged me gently but firmly towards its completion, an event that I am sure she must at times have begun to doubt. My visit to Navrongo, described in Chapter 1, was made possible through the work of Betty Kirkwood, and I am grateful to the many friends I made in and around Navrongo. Since then, Navrongo has gained mains electricity, but still lacks many important amenities. A period as a Fellow of the Institute for Advanced Study in Budapest aided the book's completion, and I thank Eörs Szathmáry and Lajos Vekas for their warm hospitality during this period. Susan Budd provided important encouragement, and a class taught by Dorothy Banks helped my writing style. Irene Johnson gave expert secretarial assistance. My editors Toby Mundy, Kirk Jensen and Ravi Mirchandani provided much appreciated advice and encouragement.

The manuscript was read in whole or part by various friends, family and colleagues whose criticisms and suggestions were a very great help. Matters of style and content were debated vigorously with my brother, Dave Kirkwood, resulting in many improvements. In particular, my mother, Deborah Kirkwood, kindly agreed to read the entire first draft of the manuscript. Little did she or I realise the exceptional challenge that this would present. In the same week that the manuscript reached her, my father, her husband of 55 years, was diagnosed as having an advanced and inoperable cancer. He died 10 days later. A short while after my father's death, my mother read and commented in detail on everything I had sent her. For reasons that will be obvious as you read the book, I place particular value on her comments made at this time.

My children Sam and Daisy egged me on finally to complete the book at the expense of most of one summer holiday together. I owe them for this and for all the fun and love we have shared. The greatest debt of all is without doubt to my wife, Louise Kirkwood, who has provided the love and sustenance without which life loses so much of its meaning, and which, when we dare to look at it squarely on, makes the reality of ageing so hard to accept.

CHAPTER 1

The funeral season

Light is sweet; at the sight of the sun the eyes are glad.
However great the number of years a man may live, let
him enjoy them all, and yet remember that dark days
will be many.

Ecclesiastes 11:7–8

The journey had been bad enough – mile after mile on broken roads
through the searing heat and dusty dryness of Africa's Sahel. But
the greeting, when the Land Rover finally swung into the driveway
of Navrongo hospital, in the Upper East Region of northern Ghana,
and came to a halt by a small whitewashed building in the shadow
of a giant baobab tree, was very much worse.

'Welcome to Navrongo! The funeral season is just beginning,'
said one of the hospital staff with a wide smile as he emerged from
the building to greet us.

'Whoops, sorry, wrong place!' I wanted to say and leap back into
the Land Rover, racing away to airport, England, and the safe
enveloping greyness of a British winter's day. But I didn't, and so
began the writing of this book.

A small market town in one of the poorest regions of the world,
with no mains electricity, no running water and precious few
amenities of any kind, is not the obvious place to write a book
about ageing. I should explain that I did not choose Navrongo for
this purpose. I was there as an extra, human baggage accompanying
an epidemiological investigation of childhood diseases. Perhaps I

was also there because Africa was the land of my birth and I have felt its call ever since. The opportunity to spend three months in West Africa had presented itself, and I had eagerly agreed.

No one could deny that Navrongo was an excellent place to study childhood diseases. The local diet was deficient in basic vitamins, the temperature regularly soared above 120 degrees (and that was in the shade), the humidity was barely detectable in the dry season yet was overwhelming in the wet, the water needed to be boiled and filtered before drinking, and parasites of all kinds abounded. Clothes hung out to dry in the open needed pressing carefully with a hot iron to kill the eggs of a parasitic fly which otherwise would hatch into larvae that burrow under your skin. For a child, quite ordinary sicknesses, like diarrhoea and even colds, superimposed on a background of poverty and poor nutrition, easily started a downward spiral that in no time at all could become life threatening. One-quarter of all children born in the region died before they were 5 years old. One might reasonably think that ageing would be far from the minds of the local health authorities, but that would be wrong. It turned out that Navrongo was not such an odd place to think about ageing, as I discovered when I went shopping.

A few miles to the north of Navrongo was the Paga border post, frontier between Ghana and Burkina Faso. This frontier was crossed often, not least because on the northern side soft drinks were cheaper and more regularly available than in Navrongo. On my first journey homewards across the frontier, I found the Ghanaian border post to be staffed by the most courteous immigration officer I have ever had the pleasure to meet.

'Thank you,' I said, as he stamped my passport after careful scrutiny of my visa.

'Thank you for thanking me,' he said in reply.

I was pondering this curious reply, wondering whether to escalate the thanks and risk staying all day, when my eye was caught by a printed poster on the wall.

'Young Today – Old Tomorrow. Take Care of the Elderly,' it proclaimed.

Now this really surprised me. Not only was this particular poster side by side with another one showing how to make oral rehydration solution to prevent death from fluid loss in babies and young children with diarrhoea – surely a pressing concern in these parts – but here we were in a region of the world where the very fabric of society is woven with the weft of respect for the elderly. There are many tribes in this part of West Africa and their linguistic and cultural roots are extremely diverse, but one thing they all have in common is that the older you get, the higher up the social ladder you climb.

I had had first-hand experience of this just a few days earlier. Driving past a small village I stopped to take a closer look at a mud-walled household compound that was painted in a particularly striking pattern of black and white stripes. As I climbed out of the Land Rover, a small crowd quickly gathered and I asked if I might take a photograph of the building that had caught my interest.

'You must ask the Old Man,' said a lanky youth whose tattered T-shirt improbably displayed the logo of the Hard Rock Café.

Looking about the crowd, I spied the oldest-looking man, who to my eye qualified as seriously venerable, and advanced to ask his permission.

'Oh no,' he said, clearly embarrassed by my gaffe. 'The Old Man is over there by that tree. Come with me.'

We walked to the shade of a gnarled thorn tree where half-sitting, half-lying was a man who seemed as old as the hills. In all likelihood he was aged no more than 80, but the climate of the Sahel is not kind to the eyes and skin, and dentists are as scarce as snowmen. Permission was asked, a small gift of money offered and graciously accepted, and I was free to look around.

You will understand, then, that I was puzzled by the poster. Back home in England, as in much of the developed world, the elderly get short shrift from the young. 'Buzz off, you old git!' is the sort of cheery greeting an old-age pensioner might expect if he remonstrates with the youth who has just kicked a football in his direction. I could see the point of a poster in my home town exhorting the youth of today to remember that they will be the senior citizens of tomorrow. But in Ghana?

The truth is, of course, that the world is changing, and changing fast. A demographic revolution is happening, the like of which has never been seen before. Nowhere is immune to it, and in some ways it is the traditional gerontocracies, like the West African tribal structures, that face the greatest strains.

The trouble is that a system that puts status and power in the hands of the old, *just because they are old*, works fine when getting to be really old is a rare occurrence. Not only are the numbers of very old people small enough to make the system workable, but there is a sound logic in trusting to the authority of someone who has been canny and robust enough to outlive his or her peers. (Although West African gerontocracies are generally male-dominated, old women also enjoy enhanced social status.) The long-lived are custodians of the tribal wisdom. For instance, they remember the terrible famine of 50 years ago and how it was survived. I am indebted to a distinguished French anthropologist with whom I once shared a somewhat boozy flight for a repellent story, but one sworn to be true, of how a village in Cameroon was saved by an old woman's memories of the way that human faeces were picked over for grains that had been only partly digested and could be washed, cooked and chewed once again for meagre but invaluable sustenance. But the system cannot work as well, if it can work at all, when the old cease to be an élite.

Over the last hundred years a revolution of unprecedented scale has been taking place in the balance of generations. The revolution began in the developed countries that were first to experience industrialisation, economic development and the health-preserving benefits of sanitation and clean drinking water. It has gained pace and is spreading fast.

In nineteenth-century England, survival to old age was a luxury that relatively few could anticipate. The following article from an English newspaper of 14 January 1885 describes the situation:

The Registrar-General has recently published the march of a generation through life in England. He says that of a million persons born
alive, the number at the end of five years will be 736,818. At the end
of 25 years there will be 684,054 of the million left. At the end of 35
years there will be 568,993. When 45 years have passed, 502,915 will
remain. At 65, 309,020 will still be alive. When 75 years have rolled
by, 161,164 (or nearly one out of six) will still remain. At 85, only
38,575 will survive. At 95, the million will be reduced to 2,153. The
number who cross the line of the century will be 223, and at 108
years from the starting point the last one will be in his grave.

Of course, the last one in his grave was probably a woman, since
women generally live longer than men, but, as we shall see in later
chapters, that is another story.

We are separated by just four generations from the date of this
article, but how the picture has changed. Let us compare the figures
from the 1880s with those for the 1990s. We will use the numbers
for both sexes combined and assume that the figures in the
newspaper article followed usual practice and lumped England and
Wales together. In each case, we will track the number of survivors
from a million newborn (see Table 1.1).

The change in survival patterns is astonishing. It is worth
looking closely at what these figures tell us. First of all, it is clear
that England and Wales have got a good deal greyer. Out of a
million people born today, five out of every six of them will still be
alive at age 65. At current birth rates, by early in the twenty-first
century one in five of the population will be aged 65 or older.

The second dramatic alteration is that life expectancy at birth in
England and Wales has nearly doubled from some 46 years in the
1880s to around 76 years in the 1990s. Herein lies a potential
source of confusion. It is not the case that *each* person in England
and Wales is now living twice as long as his or her Victorian
forebears. Life expectancy has doubled because many fewer people
are dying young.

Table 1.1

Survival in England and Wales in the 1880s and 1990s

	1880s	1990s
Born alive	1,000,000	1,000,000
Alive at 5 years	736,818	991,350
Alive at 25 years	684,054	984,230
Alive at 35 years	568,993	977,500
Alive at 45 years	502,915	963,960
Alive at 65 years	309,020	830,990
Alive at 75 years	161,164	612,740
Alive at 85 years	38,575	286,950
Alive at 95 years	2,153	45,450
Alive at 100 years	223	8,710

SOURCE: Figures for the 1880s are from the 1885 newspaper account of the Registrar-General's report. Figures for the 1990s were kindly provided by A.R. Thatcher, former Registrar-General of England and Wales.

The bottom half of our table reveals what has happened to the survival rates of the older age groups, which show a significant but more modest improvement. We can calculate the survival rates at these older ages by doing a little arithmetic. For a present-day inhabitant of England and Wales, the chance of surviving to age 75, given that you have already made it to 65, is 612,740 divided by 830,990, or 73 per cent. For a Victorian, the same chance was 161,164 divided by 309,020, which was, even then, a respectable 52 per cent. Indeed, if we measure the further life expectancy of 65-year-olds, we find only a modest increase from around 11 years in the 1880s to around 16 years in the 1990s. The ageing process now is not so very different from what it was then. Even in Victorian times there were those who lived to be a hundred.

The biggest change since the late nineteenth century is revealed in the common causes of death. The most frequent killer in the developed world today is cardiovascular disease – disease of the blood circulatory system – which accounts for 40–50 per cent of deaths in many first-world countries.[1] Most cardiovascular disease is heart disease leading to heart attacks, although circulatory disease of the brain causing stroke or vascular dementia is also of major importance. Next on the list comes cancer, which in its variety of forms accounts for another 30–40 per cent of deaths. After cardiovascular disease and cancer come accidents, of which car accidents account for about half. Accidents make up around 4 per cent of the total. Chest diseases like emphysema and pneumonia then head a cast of more minor but still important players, including diabetes, liver disease and suicide. Some of the major afflictions of old age, like Alzheimer's disease, do not feature on the list in their own right because they rarely cause death directly; they lurk behind the other killers as contributing factors.

Turn back the clock 100 years, take away our antibiotics and vaccines, and the cast of killers was a very different affair. Top was respiratory disease, not including tuberculosis and influenza. Next were the so-called digestive diseases – scourges like typhoid, cholera, dysentery and a host of other ailments. Tuberculosis was a major cause of death in its own right, as was influenza. Diphtheria, croup, measles, whooping cough and scarlet fever were all diseases to be feared, causing many deaths in childhood. And accidents, in the absence of the modern emergency room and with much greater risk of sepsis, were twice the cause of death they now are, even though our love affair with the motor car was then just a twinkle in an occasional eye.

In one sense, travelling to Navrongo was like travelling backwards in time. Mortality rates in the region around modern-day Navrongo are not so very different from those in Victorian England and Wales. A quarter of children die in the first 5 years of life;

infectious diseases are still a major cause of death; and accidents, childbirth and malnutrition continue to exact a heavy toll. But in another sense, Navrongo is very different from the England and Wales of the 1880s. The technologies of modern medicine, agriculture, industry and water purification are not waiting to be discovered at some uncertain date in the future. They are with us now and the obstacles to their wider application are economic, social and political.

Demographic change is coming to the third world even faster than it has come to the first. It is coming partly because public health is improving, even in underdeveloped countries, except where progress is undermined by the scourge of AIDS, but also because birth rates are declining. With fewer children being born, and the rest getting older all the time, the average age of the population is going up and up. By the year 2020 – just a generation away – more than a billion people in the world will be 60 or older, and more than two-thirds of them will be living in developing countries. The whole globe is getting greyer.

There is an unfortunate tendency to see the greying of the world's population as a disaster in the making instead of the twofold triumph that it really is. Firstly, we have managed – not a moment too soon – to begin to bring soaring population growth under control. Secondly, we have succeeded – through vaccination, antibiotics, sanitation, nutrition, education and so on – in bringing death rates down. If it turns out now that we lack the will and strategies to accommodate the elderly people that result from these successes, and to realise their potential as a benefit not a burden, then perhaps we should seriously question whether as a species we can justly continue to conduct our affairs under the grandiose title of *Homo sapiens*.

Coded in the statistics of life and death of the 1880s and 1990s is a message that hints at a profound shift in the psyche of most of the better-off nations of the world: we have lost our intimate familiarity with death.

In the England and Wales of Queen Victoria's day, one in four children died before its fifth birthday. One in fourteen youngsters died between 5 and 25. One in six young adults died between 25 and 35, accidents, infections and childbirth being the major killers. One in nine died between 35 and 45. And one in three died in middle age between 45 and 65. I came to appreciate the enormity of this change during my time in Navrongo.

The funeral season, it turned out, was not the dread prospect it seemed at first. Funerals in Navrongo are not the same as burials, which are simple interments and happen quickly after a person dies. Funerals are ceremonial rites of passage, held months or even years after death. They are organised at great expense to show respect to the spirit of the dead person and to facilitate his or her advancement in the spirit world, in the hope that this will also secure the ancestral spirit's protection and beneficence towards its surviving kin. Funerals are great parties with much eating and drinking that last several days and that draw family and friends from far and wide. There are even special funeral pots of baked clay on sale in the markets, just to hold all the extra food and millet beer. Because funerals are such big affairs, the funeral season happens when agriculture is at a low ebb in the drought months from January to March. Deaths happen all the time.

And death did happen all the time in Navrongo, mostly simple, preventable deaths – in childbirth or from malaria – that struck down people of any age. Sadly, this included several people within our circle of acquaintance, including a colleague and good friend who returned home to London in apparent health, only to collapse and suffer brain death from an attack of cerebral malaria. This within a small community in a few short months.

This brings me to the sombre side of this opening chapter. A book about ageing is a book that must confront death. Confronting death is something for which, in our privileged world, we are increasingly unprepared psychologically. Death strikes rarely enough that we can deny it to an extent that would seem strange in Navrongo. In the developed world, the physical aspects of dying are increasingly unfamiliar. Deaths mostly happen in hospitals, often with no close kin at the bedside. The time-honoured practice of paying last respects to the corpse has been abandoned in many sections of society. It is small wonder that we have become less comfortable with the whole idea of dying and that talking about it openly has become a hard thing to do. People who have to confront death directly – for example, those with terminal illness or the very old – do so with the added burden of so much denial around them.

The things we hide by denial in our minds have a habit of springing out and ambushing us when we least expect them. When Diana, Princess of Wales, died in a car accident in Paris in the small hours of 31 August 1997, the traditionally reserved British public surprised themselves by the strength of their emotional reaction. Of course Diana was popular, and of course her death was a tragic loss, but it was not just grief that struck so hard. Shock and disbelief were quite as evident as grief – shock at the death of one so full of life, so young. Diana's death was a reminder to us all of what we would much rather forget: our own mortality. Yet by Diana's age a century ago, nearly half of those born in the same year as her would have been dead already.

For most of us today, mortality and ageing are closely inter-twined. We don't relish an early death, but we hate the thought of growing old. When we open our minds to the reality of ageing we feel anguish. No one who is in good health, of sound mind, and capable of giving and receiving love, including the proper love of oneself, could feel anything less at the realisation that ageing will rob one's loved ones and oneself of vitality and life. But anguish, like joy, is too strong an emotion to be sustained for long. There is good news on the horizon and a challenge to be met.

The good news is that old people are living longer, healthier lives, and that the remarkable improvements in survival are continuing. The 85+ age group is currently the fastest-growing sector of the population in most of the developed countries. There are many who fear that longer lives will mean more ill health and disability, yet there is encouraging evidence that in the United States the period of sickness and disability is actually getting shorter, not longer, as life span increases.

The challenge is to age as successfully as we can. For society, the challenge of successful ageing is a paramount issue touching on all aspects of life – social, economic, medical and spiritual. For the individual, the challenge is to reach old age in optimum health and to develop the resources and attitudes to preserve independence and quality of life for as long as possible. To meet these challenges, we need to find out as much as we can about the ageing process. The following chapters will reveal what scientists already understand about why ageing happens and what causes it.

CHAPTER 2

Attitudes to ageing

An aged man is but a paltry thing,
A tattered coat upon a stick, unless
Soul clap its hands and sing, and louder sing.

W.B. Yeats, 'Sailing to Byzantium'

It's not that I'm afraid to die. I just don't
want to be there when it happens.

Woody Allen

'And they lived happily ever after,' end the fairy tales. Who are they kidding? Children are quick to learn about ageing and death, and they are quick, too, to realise that there is something taboo about the real ending to the story. Here is an alternative version of the standard fairy tale ending: 'And they lived happily for many years, until they grew old and sick together and in the end, not without pain and certainly not without fear, but smiling, they died of old age on the very same day.'

Like it? I do. And if it opens up a meaningful dialogue with our children, so much the better. Sex education has advanced in leaps and bounds. Death education lags well behind. We need it. We are growing to fear death more because we see it – real death, that is, not Hollywood fantasy death – a lot less.

I should like to tell you about two deaths that touch on ageing in different ways. The first was the death of Mr Hoddle. Mr Hoddle and I got acquainted, as often happens in England, over a cup of tea. This was not the usual cup of tea, served hot in a china cup with handle and saucer. It was a two-handled plastic beaker with a long spout, through which Mr Hoddle could suck a luke-warm milky brew, sweetened with the regulation two spoons of sugar. You see, Mr Hoddle was 92, a long-stay patient in the local National Health Service hospital, while I was a student working as a hospital domestic to earn money for a trip to France in my summer vacation.

'Cup of tea, Mr Hoddle?' I called gently as he lay with his eyes closed, head propped on a pile of white pillows. One eye opened and closed again, and then after some time both eyes opened together and gazed unblinking at me and my tea trolley.

'Cup of tea?' I asked again. With the faintest of nods Mr Hoddle showed that he had heard and understood. I busied myself with the teapot and brought the beaker carefully to the side of his bed. I moved the beaker towards Mr Hoddle's slightly shaky hands, ready to help if help was needed. Suddenly, Mr Hoddle grabbed at the wrist of my arm that was holding his tea and squeezed tightly.

'Steady, Mr Hoddle, you'll spill your tea,' I protested, struggling to keep the beaker from slopping its contents over the sheets. The grip tightened, then quickly relaxed.

Mr Hoddle leaned forward and inspected the four white bands on my wrist where his bony fingers had squeezed the blood away. His face lit into a wheezy chuckle.

The same thing happened the next day, and the next, so that it came to be a regular private ritual between us. Mr Hoddle never explained why he did this. Indeed, he talked very little.

One morning, as I turned my tea trolley in at Mr Hoddle's door, I saw that his bed was stripped and bare. His one visible possession, a photograph of himself as a young man, with boxing gloves dangling around his neck and a sports trophy in his hands, had disappeared. I asked a nurse where Mr Hoddle had gone and was told that he had died in the early hours of that morning.

'What did he die of?' I asked. 'He was fine yesterday.' The nurse reacted in the brusque way I noticed was common when questions were asked outside the line of duty. 'Old age,' she said sharply. 'He was 92.'

Even after 30 years, I remember Mr Hoddle clearly and I have often tried to guess the real point of his game. The likeliest answer always seemed that he knew his strength was failing and that squeezing the blood from my wrists gave some comfort that it had not all gone. Mr Hoddle had been tough, a fighter, and I suspect he hated to be bedridden and dependent on the well-meaning, but ultimately humiliating ministrations of the hospital staff.

Lately, though, I have wondered if Mr Hoddle was not just lonely and afraid, seeking human contact of a kind. He never had visitors and the nurses and doctors were too brisk and busy. For good professional reasons, medical staff need to keep an emotional distance from their patients. If this was the case, then I regret that I could not have been of more help to Mr Hoddle, but it makes me glad of our game.

The second death happened on the road to Navrongo. It was the death of a kid – a young goat, that is, not a child. But children have a role to play in this story, because they were a part of our party.

The kid ran in front of the Land Rover at the last possible moment. The driver, a local man, had no time to react before the head struck the front of the vehicle. In a simple, eloquent gesture, he beat one of his hands on the other forearm to signal regret to the villagers who stood by the road, and received a wave of acknowledgement. He did not stop. There was no point.

'Dad, what was that noise?' piped up a small voice from the back seat.

'We hit a goat,' I answered.

'Is it dead?' asked the small voice.

'I think so.'

At this point a second small voice began a whispered conversation with the first small voice.

'Was it an old goat?' asked the second small voice.

'No, a young one,' I answered. 'Why do you ask?'

More whispering.

'Only if it was an old goat it wouldn't have mattered so much because it wouldn't have had so long to live anyway,' said the second small voice.

Here we come to one of the trickier aspects of the whole business of ageing. How do we gain a proper view of something so deeply ingrained in our consciousness that we have formed all kinds of preconceptions and prejudices about it?

Two attitudes in particular have been responsible for more distortions in our understanding of ageing than any others. These are *fatalism* and *ageism*. Fatalism leads to accepting that ageing must happen because it is part of the natural order of things to wear out. It results in too ready an acceptance of limitations imposed on our daily lives as a result of changes in our bodies as we get older. Certainly it is true that we cannot run as fast at 50 as at 20, and at 80 it may not be sensible to run at all. However, as 80-year-olds we should not just curl up and wait to die. A few years ago an acquaintance, then in his early sixties after a career of demanding physical work, went to see his medical practitioner because of recurring pain in his shoulder.

'Wear and tear,' diagnosed the doctor. 'What do you expect at your age? Nothing to be done about that.'

But the acquaintance sought a second opinion, received a straightforward course of physiotherapy, and has suffered no pain in the shoulder since. The first doctor should have known better, but I doubt if there was malice in his negligence. He merely fell into the trap of fatalism which is there to ensnare us all.

Fatalism also intrudes its distorting influence into our perception of ageing as a general process in biology: 'Death takes place because a worn-out tissue cannot forever renew itself,' wrote the German naturalist August Weismann, who in 1881 offered the first detailed theory of why ageing happens. The fatalistic fallacy embodied in this apparently circular argument can be found again and again in writings on ageing, and has led to a particularly damaging set of preconceptions, as we shall see in later chapters.

Weismann came to set great store by the difference he identified between the 'immortality' of reproductive cells, the cells in the body that carry genes forward to the next generation, and the 'mortality' of the rest, which will age and die. Four decades later, Sigmund Freud in his pivotal work *Beyond the Pleasure Principle* would draw heavily on Weismann's ideas as he sought a biological basis for the distinction between what he termed *thanatos* (the death instinct) and *eros* (the life instinct) of the organism.

'What lives, wants to die again. Originating in dust, it wants to be dust again. Not only the life-drive is in them, but the death-drive as well,' wrote Freud. However, the idea of a universal death-drive, with its fatalistic undertones, finds scant support in modern biology (see Chapter 5).

Ageism is the devaluing of an individual solely because of his or her age. This is as foolish as it is unfair. Some of the worst extremes are seen in the infantilising and occasionally brutal treatment of older people in institutional care and even in the family home. Like child abuse, but with the generations reversed, ill-treatment of older people in the family home can pass unrecognised because its very existence strikes at the heart of our notions of what families are about. Ageism is a distortion to rank with racism and sexism in its pervasiveness of society.

As with sexism and racism, extreme instances of ageism can be easily recognised and dealt with, provided the motivation exists to do so. It is the subtler, more deeply ingrained attitudes that require sensitive self-examination and even training before we are aware of them. They are there nonetheless, in varying forms and to varying degrees, and they have a powerful hold on us. The old, like the disabled, may find themselves being talked about ('Does she take sugar in her tea?') rather than addressed directly, or they may find well-meaning but incorrect assumptions being made about them ('Oh no, I don't think he'd enjoy that at his age'). We need to understand that frailty and physical dependence do not by themselves imply weakness of mind and spirit.

The fast pace of change in present-day society adds to the

disadvantage experienced by the old. For example, a 70-year-old in the 1990s learnt his or her vocabulary and manners in the 1920s and 1930s. Small wonder that the old may have difficulty understanding the teenage assistant in the supermarket. Confusion can arise from many sources, but we are all too ready to assume a senile confusional state in an old person, when the problem is often nothing more than simple unfamiliarity with a changing world.

One particular feature of ageism sets it apart from most other types of prejudice: we are all its potential victims. The pre-judgers and the pre-judged are not separated into their camps at birth, but are distinguished from each other by the passage of time. The pre-judgers *become* the pre-judged, again reversing the pattern so often seen with child abuse, where the abused become abusers in turn. The pre-judgers carry with them the prejudice that subtly turns against themselves.

A good way to test how we value people of different kinds is to put ourselves in the imaginary position of making life-and-death judgements about them. Imagine that you are the pilot of a small helicopter on a remote island from which a party of six holiday-makers has set sail in a boat equipped with a radio. Suddenly you receive a mayday call to say that the boat has met with an accident and is sinking slowly in seas too cold for survival. The boat is 20 minutes flying time away, you can rescue only one person at a time, and no other rescuer is on hand. The party on the boat cannot agree on the order in which they are to be rescued and ask you to make this decision for them. They will abide by your decision without question. You have 20 minutes to decide. You know that in all probability you will not be able to save all of them.

You have spent time with the group of holiday-makers on the island and know them all well, but have no particular friendship with any of them.

Eric, 64, is the leader of the group. He worked hard in insurance since leaving school and is looking forward to retirement in a year's time when he plans to buy his own boat and sail it around the

Caribbean, a lifelong dream. Eric is a bachelor and enjoys his solitary existence.

Jane, 33, has been a teacher and plans to return to work when her children, now 2 and 4 years old, begin school. The children are with their father, who is keen that Jane should have a real break from domestic chores, and who has taken holiday from his job as a supermarket manager to make this possible. Jane is anxious about how he will be coping while she is away and feels a little guilty about enjoying herself.

Wayne, 42, is finance officer in a worker's co-operative that was established to save the jobs of 250 metalworkers when the parent company decided to close the works down. It is largely due to Wayne's shrewd business sense that the co-operative has been a success. He has three children, aged 16, 14 and 11. His wife has a well-paid job as a personnel officer in local government.

Jessie, 18, has just left school and intends to continue her studies at college after she has taken a year to 'discover herself'. She is a popular young woman and has unusual self-confidence for her age. She has no fixed plans, but knows vaguely that she would like to do something useful.

Jeff, 25, is owner of the boat and has never worked. He is wealthy from insurance money he got when his parents died in a plane crash 10 years ago. It was Eric, incidentally, who arranged the insurance policy and who tries, unsuccessfully, to advise Jeff in investing it. Jeff is single, races motorbikes semi-professionally, and has frequent casual affairs with women. He has told you in confidence that he has two illegitimate children for whom he pays generous, but unofficial maintenance.

Constance, 78, is recently widowed. After 2 years devoted to the care of her husband during his slow and difficult terminal illness, she feels liberated to take up new interests. Her four grown-up children are rather horrified at her 'foolhardy' decision to join this adventure holiday. Her seven grandchildren, on the other hand, are thrilled at the idea and she has been writing a diary about it for them.

There they are and their lives are in your hands. The clock is running and you must decide. Why not put this book down for a few minutes and think about it?

Time is up, the boat is in sight, and you must give your instructions about who is first to be taken off.

Several factors probably contributed to your thoughts as you weighed one life against another. You are a decent person and you have tried to be as objective and fair as possible in almost impossible circumstances. Jane, Wayne and Jeff all have dependent children. Wayne has colleagues whose livelihood he secures. Jessie has all her adult life before her. Eric and Constance deserve the new opportunities before them after lives of hard work.

Of the things you know about them, you are likely to take particular account of age. When disaster strikes we tend automatically to think of years lost, as with the goat on the road to Navrongo. Age counts against Constance and Eric, and in favour of Jessie and Jeff.

Placing a premium on youth is not intrinsically ageist. Many see this as simple realism. Life insurance companies do it all the time. Newspapers feed us age data in nearly every item of news. Conventional wisdom tells us we *need* this information to weigh the significance of the news. Sometimes we do. But we should be aware of the value judgements we are making, for these are slippery issues, and value is in the eye of the valuer. Later in this book we will look more closely at the value that organisms place on their survival in the natural world.

In their plight on the ocean wave, one possibility has not crossed the minds of Eric, Jane, Wayne, Jessie, Jeff and Constance, but might occur to you, especially if you happen to find that in the pocket of your flying jacket there is a small die. Let chance decide.

By throwing the die you are absolved from individual responsibility for their fates. Not only this, but you might even help the relatives of those you cannot rescue. It would be painful for them to know that their loved one perished because he or she was not thought important enough. It might be easier if death were the result of random accident.

The drawback to using the die is that even this is not entirely fair. For Constance the average risk of dying on any given day, taking account of her age, is about 180 times as great as for Jessie. The others fall in between, though Jeff has a riskier lifestyle than most. You could be accused of practising a scheme that is disadvantageous to the young.

We will not worry about the finer details of the problem (but remember it is a situation faced daily by doctors who must decide, for example, which patients are highest priority for a life-saving kidney transplant). There is no right answer. Each of us will like one scheme better than another. The point is to ask ourselves why.

Let us stay with the idea of an equal chance of survival a moment longer, for when applied to all the hazards of life, this is the goal of those who would seek to banish ageing altogether. Absence of ageing is not absence of death, which in the material world is a physical impossibility. Not to age is to be in a steady state where your risk of dying does not increase from one day to the next. The ancient alchemists sought the secret of such immortality in their quest for the elixir of life; it remains for some a seductive goal. A Texan millionaire, for example, believes that abolition of ageing is both feasible and desirable and has offered a surprisingly modest prize for the scientist who discovers the 'cure'. Imagine a world where your future life expectancy is constant, no matter how long you have lived already, and depends only upon the level of risk in the lifestyle you have adopted. What potential to develop your pleasures and skills! But what a calamity a death would be. I fear we might all become excessively cautious.

A lesser, but more realistic goal than that of the immortalists is to extend human life span. The issue of whether or not it is a good idea to extend our life span generates a good deal of discussion, even alarm, among responsible members of a species already concerned about global overpopulation and environmental pollution. Nevertheless, medical researchers are already doing life extension research as they seek ways to prevent and treat heart disease, cancer and the other major killers whose incidence increases so steeply as we get older. To recognise that these diseases may share some of the basic cellular and molecular processes that are responsible for old age is to recognise the true goal that most biological gerontologists aspire to. No one who has experienced at close hand the course of senile dementia, such as is caused by Alzheimer's disease, will deny that there is a desperate need to uncover the causes of debilitation which leave an older person still alive, but cripplingly dependent on those who care for him or her.

I am sure that as the pilot of the helicopter you would, if you could, have saved the lives of all six holiday-makers in the sinking ship, regardless of the fact that Constance was old already. I know that I would have done, and that Constance's obvious enjoyment of life would for me have counted in favour of her early rescue. But in terms of ageing, we are all, metaphorically speaking, in sinking ships. So if we question the wisdom of extending life by fighting ageing, but not by rescuing holiday-makers from drowning, we need to agree just when and why extra life is not worth having. And most important, we need to agree who is to decide. As a gerontologist, I am sometimes asked how long I would personally like to live. My answer is this: I want to live as long as my quality of life is good and I can look forward to each new day.

CHAPTER 3

What's in a name?

Sweet rose, whose hue angry and brave
Bids the rash gazer wipe his eye:
Thy root is ever in its grave,
And thou must die.

George Herbert, 'Virtue'

So Mr Hoddle died of old age? No doubt the nurse was right, but I doubt if anybody wrote 'old age' as the cause of death on Mr Hoddle's death certificate. The International Classification of Diseases does allow for old age – or, technically, senility without mention of psychosis – to be recorded, but less than 1 per cent of death certificates in the UK show this cause, and less than one-tenth of 1 per cent in the USA. Death certificates tend to record specific diseases, like pneumonia or coronary thrombosis. The death certificate does, however, provide space to list contributory or secondary factors. In the case of a very old person, this list might be long. Death from old age means, in effect, that a person's hold on life has become so precarious that, had it not been this particular cause of death today, it would have been another tomorrow.

The situation is rather like an old automobile that has come at last to its final breakdown. The end might equally well be due to engine failure or to the rusting through of a vital body part. The result is the same. The car is broken and will run no further. But is this what ageing really means?

'Ageing' is a word that we use all the time, so we have a fair idea what we mean by it in general terms, but it is when we try to be a little more precise that the trouble starts. For example, within a population there are some individuals who seem to age fast, and others who age more slowly. To describe this apparent variation in the rate of ageing, scientists like to draw a distinction between 'chronological' and 'biological' age. Chronological age is measured by the simple passage of time – the number of candles on a birthday cake. Biological age tries to express how far we have travelled along the road from birth to death of old age. A woman who is aged 60 years chronologically, but has the appearance and stamina of a typical 50-year-old, may have a slower biological ageing rate than the average. She might expect to live longer. We all know of people who are 'remarkably well preserved' for their years, and others who have 'aged before their time'. But biological age is not so simple a concept to measure, and in medicine and biology the definition of what ageing really is has vexed the experts for many years.

One of the trickier questions is the relationship between so-called 'normal' ageing and the various age-related diseases, like osteoporosis and cataract, which affect more and more people as they grow older. Are such diseases *part* of the ageing process, or can we only describe what happens in disease-free individuals as normal ageing?

Blood pressure varies from individual to individual but, on the average, blood pressure rises with age. By convention, people with blood pressure higher than a certain cut-off value are labelled 'hypertensive' and may receive treatment. A greater fraction of the population is hypertensive among older age groups. If we want to describe normal ageing, should we restrict our attention to the shrinking fraction of people with 'normal' blood pressure? Or does this distort our picture of normal ageing by excluding a fraction of perfectly typical old people whose blood pressure exceeds what is, after all, an arbitrary cut-off?

Grasping the correct distinction between normal ageing and disease smacks of a semantic quibble, but words are powerful, and the consequences of how we use them can be far-reaching. John Grimley Evans, professor of clinical geratology at the University of Oxford, puts the matter well:

> At a practical level 'disease' is accepted as comprising phenomena that are the proper and indeed obligatory concern of the doctor, and a 'disease' entitles its victim to sympathy and social support, subject to the constraints of his fulfilling the sundry obligations of the 'sick role'. 'Ageing', to the common mind, is the universal ineluctable and unameliorable lot of mankind, entitling its victim to no particular sympathy and not qualifying him or her for medical attention. Indeed, those victims who attempt to conceal or prevent the ravages of ageing may bring onto their heads the obloquy reserved for those who exhibit pretensions to qualities – in this case, youth – that they do not possess. Those who struggle to overcome the impairments of disease, in contrast, are admired for their courage in adversity.

A good example of how diseases are treated as separate from the normal ageing process can be seen in the spectacular rise of Alzheimer's disease in recent decades as a research priority, in terms both of the money that is spent on it and the scientific cachet that attaches to its researchers. When Alzheimer's disease was merely a disorder under the general heading of senile dementia, it had a low public profile. But once the medical profession had recognized it generally as a disease, it gained a different status. The victims of this unfortunate malady suffer as before, but their affliction elicits greater sympathy. And for a *disease*, there is the hope of a *cure*. Through the investment of much outstanding research effort, the prospect of understanding the molecular basis of Alzheimer's disease is now bright (see Chapter 9). From this knowledge may come effective therapy.

Is this revised perception of Alzheimer's disease a model of how we should tackle the other problems of old age, or does this duck the real issue of getting to grips with the basic processes of ageing, some of which may underlie Alzheimer's disease itself?

Most medicine in the developed world is concerned with diseases that become more common with age. The leading causes of death in industrialised nations – heart disease, cancer and stroke – are all diseases that are rare among young adults, but strike with ever greater frequency as we get older. In countries such as the United States, heart disease, which is currently the greatest killer, typically carries off each year around one in every forty 65–69-year-olds, one in twenty-seven 70–74-year-olds, one in seventeen 75–79-year-olds, one in eleven 80–84-year-olds, and one in seven 85-year-olds and over.

Heart disease and cancer are intriguing examples of conditions that occur more frequently with ageing, but they need not be as closely related to ageing as they seem. Richard Peto, cancer epidemiologist at the University of Oxford, some years ago expressed an extreme view: 'There is no such thing as ageing, and cancer is not associated with it,' he wrote.

Peto's provocative statement was intended to challenge the idea that old people get more cancer because aged cells are intrinsically more likely to become malignant on being exposed to a cancer-causing agent. What Peto argued instead was that old cells have just been around for longer, so their cumulative exposure to cancer-causing agents has been that much greater.

The basis of Peto's claim was an experiment in which a chemical called benzpyrene, which is known to cause skin cancer, was painted on to the backs of mice of different ages and left for differing amounts of time. Now if ageing makes cells intrinsically more cancer-prone, then the same duration of treatment applied to older mice should produce more cancers than in younger mice. On the other hand, if it is just the duration of exposure to the cancer-causing agent that results in the rise in cancer incidence, then leaving benzpyrene on the backs of the mice for a given period of time should produce the same increase in cancer incidence, regardless of whether the mice were old or young at the start of the experiment.

When the results of the experiment were examined, it was the second of these predictions that was found to fit the facts. In ordinary animals, not treated with agents like benzpyrene, the reason why old mice develop cancer more often than young mice is not because their cells are intrinsically more cancer-prone, Peto concluded, but because they have been exposed to the cancer-causing slings and arrows of their environment for longer.

The conclusion that ageing and cancer are unrelated is too extreme, however, because both are probably caused by similar insults to the cells of our bodies. In fact, there is growing evidence that the link between cancer and ageing is a deep one (see Chapter 10). With hindsight, the benzpyrene may have been so strong an inducer of cancers that it masked the gradual accumulation of mutations that is thought to underlie the age-associated increase in spontaneous (as opposed to experimentally induced) tumours. A strong corrosive will eat through the body shell of an old and young car at about the same rate, even though rusting in the old car will be more advanced. Peto was right, however, that we need to be rather careful about how we attribute the rising incidence of diseases in older people.

If you are over 35, you will have some experience of ageing already – the odd grey hair, the occasional wrinkle, slightly less turn of speed as you run for the bus. If you are older, your sense of change will be greater. We all know these signs of ageing and can recognise them instantly in others, so that on the whole we can place a stranger's age with a reasonable degree of accuracy. But measuring them in a systematic and scientific way is another matter. In spite of the fact that ageing is a universal process in humans, the details of how individuals age vary considerably. When this variation is added to the variability with which we begin our lives, it proves surprisingly hard to define how ageing affects us in ways that allow our biological age to be measured.

In 1958 at the Gerontology Research Centre in Baltimore, Maryland, later to become a part of the US National Institute of Ageing, an ambitious project to measure how people actually age was begun. The famous Baltimore Longitudinal Study of Ageing, founded by physiologist Nathan Shock, enrolled a group of volunteer men who at the time of joining the study were healthy, successful members of the community. Initially, the study was modest in size, but as resources expanded, more subjects were admitted and about 650 men, ranging in age from 18 to about 100, have been regularly tested. Since 1978, women have also been included.

The primary aim of the Baltimore Longitudinal Study has been to examine the effects of ageing in healthy men and women who have not suffered the disadvantages of poor education and low or marginal incomes. The subjects of the study are a self-selected and privileged group, and only those who remain free from evidence of major disease are included in the analysis of age differences and trends. With these restrictions the study is obviously not representative of the population at large, but two justifications can be offered for the selection criteria that were used. First, there are many ways in which poor housing and diet can shorten life, but these do not necessarily have anything to do with ageing. Second, the Baltimore study requires an exceptional degree of commitment from its participants, who report every 18–24 months, travelling at their own expense, to undergo a 2–3-day battery of clinical, physiological, biochemical and psychological tests. To go through all this you have to really want to do it.

Most of the subjects were encouraged to join the study by other participants, particularly from among residents of Scientists' Cliffs, a community on the western shore of Chesapeake Bay about 60 miles south-east of Baltimore. Many are interested in scientific aspects of the research, as well as in their own ageing.

So after all the hours on the treadmill, the X-rays, the intravenous glucose tolerance tests, echocardiograms, eye tests, memory tests, blood chemistry tests, filling out of diet diaries and psychological questionnaires, what have the enthusiastic Baltimore volunteers found out about themselves? Although the study is ongoing and the results are not yet fully analysed, one of the important things to emerge very clearly from it so far is that we do indeed age differently.

Average trends – for example, an average decline in creatinine clearance, which is an indicator of kidney function – do not give anything like a full picture of what may be happening inside you and me. Mr and Ms Average are the exception. Just as most people are taller or shorter than the average height in the population, there are those whose creatinine clearance goes down more steeply with age than the average, while there are others for whom it even goes up!

The personality tests in the Baltimore Longitudinal Study are revealing for their absence of change. Adult personality remains very stable, with only two characteristics showing any effects of ageing. No one is likely to be surprised that preference for fast-paced activities starts to decline at age 40. And among men, traditional masculine interests, such as an enthusiasm for sport, heavy drinking and taking engines apart, show a decline over the total age span from 30 onwards. Like creatinine clearance, individual differences in personality traits are far more pronounced than age trends, and no evidence has been found for consistent changes with age.

Contrary to popular prejudice, the Baltimore study suggests that old people do not become preoccupied with concerns for their health. A study of 70-year-olds in Göteborg, Sweden, has shown that underdiagnosis of disease in old people is common, principally because the old *expect* to have certain impairments at older ages and do not even mention them to their doctors, even though standard treaments may exist for their conditions.

Something that gerontologists, and the insurance industry, would dearly like to be able to do is to haul you in from the street, carry out a series of simple tests on you, and determine how old you are biologically. This, as we saw earlier, is different from your chronological age, which simply indicates how long has passed since your birth. If you happen to be one of life's fortunates and have aged more slowly than the average, your biological age will be less than your chronological age.

Indicators of biological age are known as biomarkers. The value of a good biomarker is that it can predict what will otherwise take a long time to measure, namely lifespan. If you want to examine the association of longevity with some intrinsic, perhaps genetic, factor, then a biomarker that gives an early indication of the rate of ageing is extremely useful. Similarly, if you want to study the effect of a drug or toxic substance in slowing down or accelerating basic ageing processes, you want a biomarker.

Many things alter progressively with ageing and any of these has the potential to serve as a biomarker, yet in spite of an extensive search the results so far are only partially satisfactory. Individual biomarkers provide some information on the overall rate of ageing, but numbers of them need to be combined in a battery of tests before the information becomes at all useful. The reason is simple: ageing is variable.

Imagine a biomarker – for example, the level of a blood enzyme – which shows on the average a steady reduction with age. Suppose the enzyme reaches its peak level at age 20 and let us say that this peak level measures, on average, 100 units of enzyme activity per millilitre of blood. Suppose also that for each additional year after age 20 the level of the enzyme falls by one unit per millilitre. Now take a sample group of 50-year-olds and measure the same biomarker in each of them. You are likely to find that the values for the biomarker among the group vary quite considerably. Since 30 years have elapsed since age 20, we know that the average will have fallen to 70 units/ml but a range of 55–85 units/ml would not be surprising.

Now imagine trying to use this information to assess the biological age of the next 50-year-old to walk through your door. Say her value for the biomarker is measured as 78 units/ml. Before you rush to decide that the woman has aged only 22 years biologically since age 20 and therefore has a biological age of just 42 years, you will need to know the following. What was her own level at age 20? How variable is the test itself? Perhaps if you retested the same woman the next week, you might get a value of 65 units/ml, suggesting a biological age of 55. And how much variation within the population is there in the rate of decline (remember creatinine clearance in the Baltimore Longitudinal Study)? Do those with faster rates of decline actually age faster, or is the clock speed of the biomarker different in different people? The catch is that to know all this you need an independent means of assessing biological age.

I do not wish to paint too bleak a picture of biomarker studies, for the work is vitally important in research and must go on. But we need to be extremely cautious of claims to be able to test biological age, which I suspect will become commoner as our collective concern about ageing grows stronger. Recently, I received a letter from a man in his early thirties who believes that he has not aged biologically since he was a teenager. Implausible as this may sound, it would be hard to prove or disprove on the basis of the tests currently available. I replied pointing this out and suggested that any test might be more conclusive when he has aged chronologically by a decade or two more. Fortunately for him, if he is correct in his belief, time is on his side.

We began this chapter by asking what we mean by 'ageing' and we have looked at the relationship between ageing and disease and at how ageing can be measured. We have seen that in spite of our general familiarity with the process, the precise concept of ageing is slippery to grasp, like a bar of soap in a bath. Just when we think we have it, it shoots out from between our fingers. We will now try to define just what ageing really is. There are two ways we might attempt this.

The first is in terms of the many changes that come about as we, and other organisms, get older. Our skin wrinkles, our bones get thinner, our senses decline, our ability to regulate body temperature is impaired, and so on. But we have seen how variable these changes can be, and more important, they really constitute *descriptions* rather than a definition.

One of our major objectives in the chapters that follow will be to understand *why* ageing happens. For this we need a definition of ageing that can be applied in species where the physical signs of ageing may be quite different from those in humans. A fruitfly does not get grey hairs, but it ages nonetheless.

The big difference between an animal that ages and one that does not is this: an animal that ages experiences a growing risk of dying as time goes by; an animal that does not age retains a constant risk of dying. But risk is not something that is easily measured in individuals. One day you are alive, the next you may be dead. How can risk be quantified from this? The answer is to define ageing and the risk of dying as attributes of a population.

When a population of mice is kept in optimum conditions in a laboratory, free from predators and infections, well fed and watered, and maintained at a comfortable temperature, it is found that initially the risk of dying, or death rate, is very low. As time progresses, the death rate steadily increases. Eventually the last mouse dies and the population is extinct. This can be represented pictorially as a graph showing the percentage of mice surviving plotted against age. Such a graph is called a survival curve (see Figure 3.1).

The survival curve of a species that experiences ageing, like the mouse, has a characteristic shape. There is often a dip down at the beginning, which represents deaths among the newborn. Then the curve enters a period when it is relatively flat. This is the period of young adulthood, when the risk of dying is low. But in middle age the survival curve turns downwards again as the risk of dying starts to climb. In the end, the curve glides down to meet the horizontal axis – in other words, survival drops to zero – at an age that corresponds to the maximum life span of the species.

The survival curve of a human population is like that of the mouse except that the flat portion spans a greater fraction of the total picture (see Figure 3.2). The agent that generates this kind of survival curve can be appreciated if one considers how the risk of dying increases with age. The picture is stark. With each extra year

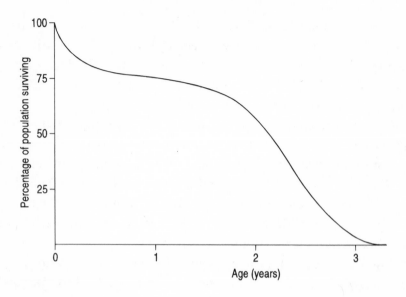

FIGURE 3.1
Typical survival curve for a population of mice kept in a protected
environment, away from the dangers of the wild. The rapid drop in
survival at young ages is due to the higher mortality of juveniles. The
curve starts to level off when the mice reach adulthood but drops again
when they become old.

of life, the risk of dying goes up, slowly at first, then faster and
faster, until around the age of 100 the risk of dying in the next year
is as high as 40 per cent.

A crumb of consolation is that, if you make it past 100, the risk
does not necessarily go up much higher. The most likely explanation for this apparent slowing down of the increase in human
mortality is that the survivors come to be less and less representative of the population at large. If there is a genetic variation for rate
of ageing (and we shall see in Chapter 14 that there appears to be),
then individuals who make it past 100 probably had a slower rise in
mortality all along. In genetic terms, the most frail types die first,
and at later ages only the most genetically robust types are left.

The rise in the adult human death rate has the convenient
mathematical property that, like money in a fixed-interest savings

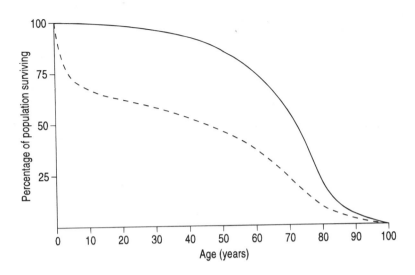

FIGURE 3.2
Survival curves of human populations under more and less favourable
conditions. The curves here are drawn for the populations of England
and Wales in the 1880s (lower, dashed curve) and 1990s (upper curve),
using the statistics given in Chapter 1.

account, it increases in an exponential way. The death rate doubles
every 8 years. This useful observation was first made in 1825 by the
English actuary Benjamin Gompertz, and it brings a nice concep-
tual simplification to the effect of ageing on mortality. Whatever
your risk of death today, 8 years hence you will be at twice the risk
of dying. And so it goes on. In 16 years, your risk will be four times
as great; in 24 years, eight times. It was on this basis that in
Chapter 2 we could assert that Constance (aged 78) was 180 times
as likely to die as Jessie (aged 18).

The Gompertz property is found in other mammals too. The
mouse has a mortality rate doubling time of about 4 months, a dog
around 3 years, a horse around 4 years. It is a neat aid to
understanding what we mean by different *rates* of ageing. But we
ought not to get too hung up on the Gompertz property. It was once
thought that organisms had to fit the Gompertz pattern as if it were

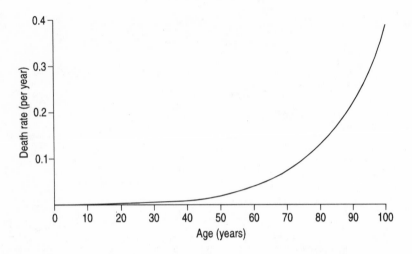

FIGURE 3.3
Human mortality rate increases with age. The curve is drawn using
data from present-day England and Wales and is typical of developed
countries.

some kind of biological law. But there is no reason why this should
be the case and clear exceptions have been found, particularly
among insects. In 1992, two major studies were reported, one on
fruitflies and the other on medflies, showing that at very old ages
the death rates of the survivors stopped increasing, and, in the case
of the medflies, even declined. Just why the death rates should stop
rising in old flies remains unclear, but there are several possible
explanations. Part of the answer may be genetic heterogeneity of
the kind just considered to explain the slowing of the death rate
increase among human centenarians. Part may be non-genetic
heterogeneity; some flies may, for all kinds of chance reasons, turn
out to be tougher than the rest. Finally, very old flies behave
differently from young flies – for example, crawling instead of
flying – which may reduce their risk of dying.

A flexible and general-purpose definition of ageing was offered
some years ago by John Maynard Smith, professor of biology at the
University of Sussex, who has a knack for putting things suc-

cinctly: 'Ageing is a progressive, generalised impairment of function resulting in an increasing probability of death.' There is, of course, a lot more to ageing than this, but as a benchmark definition it will do very nicely.

Grasping this yardstick in our hands, we can now venture forth into the world of richly varied animal (and plant) life and discover which species age and which do not. The ones that age will be those that show an increasing probability of death as they get chronologically older.

In order to do this, we need to observe a population of individuals, note when they are born and when they die, and then plot their survival curve. Our sample ought not to be too small or chance may unduly influence the result. A sample size of 100 is a reasonable minimum. Unfortunately, it very quickly becomes apparent that this is not entirely straightforward.

The first thing we need to make sure of is that a sufficiently large proportion of the sample actually live long enough for us to have some chance of observing them age. We do not want them all dying from an infection at an early age, or being eaten by predators. We need to protect them, at least partially, and this means keeping them in captivity or in a protected reserve. This costs money, and some species adapt poorly to life in captivity.

The second thing we need to do is wait. In the case of a large mammal like an elephant (which can live for 80 years), and even more so in the case of a giant tortoise (which may live 150 years plus), we may need to wait a very long time. It will come as little surprise that we know rather more about ageing in short-lived species than we do about ageing in long-lived species.

In spite of these difficulties, a fairly good picture of the prevalence of ageing among the various species can be pieced together from records of research laboratories, zoos, farms, nature reserves and some observations in the wild. Mammals, our closest relatives, nearly all seem to age in much the same way as we do ourselves.[2] Birds are believed to age without exception, generally living longer lives than comparably sized mammals. The common seagull, for instance, can live for 40 years or more, considerably longer than a cat (28 years) or a dog (20 years), and nearly as long as a gorilla (49 years). Fish, snakes, lizards, alligators, tortoises, frogs and all other amphibians and reptiles appear to age, although some do so only very slowly. Insects in their huge variety have many ways of living out their lives, some long (such as cicadas, some of which live more than a decade), some short (such as fruitflies, which live a few weeks), but all apparently finite. Insect life spans are complicated by the need to add in the larval stages (the adult forms sometimes survive only a day, as in the case of mayflies), and by the fact that, as with other cold-blooded creatures, the life span may be affected by temperature (being generally shorter at higher temperatures).

Our survey has taken in much of the animal kingdom, and we have found so far a catalogue of species all of which are believed to have finite life spans. We have not yet encountered an animal that does not age, although there is good evidence that such species do exist.

To prove the immortality of a species is, of course, impossible. You might watch your sample population for as long as you can bear and not detect a rise in the death rate, but you cannot rule out that ageing will occur a week or a year later. You cannot watch for ever. But if you are prepared to accept a verdict of 'immortal beyond reasonable doubt', then take a good look at the sea anemone, that jellylike blob of flesh crowned with a mass of waving tentacles, which is found in rockpools at the seaside.

A batch of sea anemones was collected at some unrecorded date prior to 1862 and kept in an aquarium at the University of Edinburgh for 80 to 90 years, until one day in 1942 they were all found simultaneously to be dead. During all this time the animals were examined on a regular basis and they showed no outward signs of deterioration. In particular, they were observed to carry on reproducing, which sea anemones do by budding offspring from their bodies. If ageing happens, it often shows up not only through an increasing death rate, but also through a decline in fertility. The sea anemones continued budding throughout their long lives. Why the sea anemones all died in 1942 remains a mystery, but it seems unlikely that they aged and died simultaneously with no prior sign. The more likely explanation is that their death was the result of an infection in the tank, or an accident (perhaps the wrong setting on a thermostat, or the switching off of an airpump, to which no one confessed).

There are other animals, notably the freshwater hydra, that appear to have the same power of immortality. Like sea anemones, each hydra has a tubular body, tentacles and a mouth. Pale green in colour, hydra are quite small, generally no more than a centimetre long. They are found in ponds and other still water, often attached to vegetation or to stones, and they feed on a variety of tiny aquatic animals that they catch with their stinging tentacles. Young hydra bud from the sides of the parent's body as miniature versions of the fully grown form.

One of the features of hydra is their extraordinary power of regeneration, demonstrated in a remarkable experiment in 1746 by Abraham Trembley, a Swiss scientist who was working in The Netherlands. After demonstrating that hydra could be chopped into little pieces, each of which could regenerate a whole animal, Trembley tempted a hydra to open its mouth by offering it a morsel of food. With great dexterity, Trembley then used a pig's bristle to push the base of the hydra's body out through its mouth, turning the unfortunate creature inside out. Before the hydra could recover from this insult, Trembley skewered the mouth with another, finer bristle, which prevented it from turning itself right side out again. Fascinated, Trembley observed that the inside-out hydra simply rearranged its much-abused self, expelled the fine bristle from the tissues around its mouth, and continued life as normal.

Trembley's experiment testifies to the indestructibility of the hydra, but many years were to elapse before the immortality of the hydra, in the sense of its not ageing, was seriously examined. In the early 1990s, Argentinian biologist Daniel Martínez, working first in New York and later in California, took up this challenge. During 4 years of painstaking observation, Martínez watched, and watched, and watched. Through all this time his collection of hydra, kept in a laboratory, showed no increase in mortality and no progressive decline in their rate of budding. The 4-year-old hydra were in no detectable way different from their original selves. We cannot know, of course, that a hydra does not age at 5 years, or 10, or 50, and anyone who wants to watch longer is welcome to try. Martinez has terminated his experiments because he is satisfied that the hydra is immortal. I, for one, believe him.

If we accept that hydra and sea anemones are immortal, the scientific mystery of ageing deepens. We need to explain why so many species age and die, when others do not. We shall return to these questions in Chapters 5 and 6.

CHAPTER 4

Longevity records

Methus'lah live nine hundred years,
Methus'lah live nine hundred years
But who calls dat livin'
When no gal'll give in
To no man what's nine hundred years?

Ira Gershwin, 'It Ain't Necessarily So'

The title of this chapter has two meanings, and we will examine them both. On the one hand, we have the record-breakers, the people who make it into the *Guinness Book of Records*. On the other hand, we have the accumulated statistics concerning duration of life as it is lived by the different peoples and animals who inhabit this earth. The credibility of the former kind of record depends upon the reliability of the latter. If fishermen are always to be believed, we might have some odd ideas about the sizes of fish. And so it is with the length of life.

At the time of writing, the world's longevity record for a human being is held by the late Jeanne Louise Calment, who was born on 21 February 1875 in Arles, France, and who died on 4 August 1997, at the age of 122 years and 5 months. Madame Calment lived through the rule of seventeen presidents of the French Republic and her life spanned an era of enormous change throughout the world. No electric light shone on the infant Jeanne Calment for the simple reason that the electric light bulb had not been invented, yet news of her death flashed through the myriad connections of the Internet to appear within seconds on computer screens around the world. The First World War and Russian Revolution began when Jeanne Calment was already nearing middle age. The Second World War ended when she was 70. Throughout these turbulent times, Jeanne Calment's body kept itself going – on and on and on – sustained in her view by a diet rich in olive oil, regular glasses of port wine, plenty of exercise and a good sense of humour.

How do we know this to be true? Our trust in the authenticity of Madame Calment's claim is based on the careful researches of French epidemiologist Jean-Marie Robine and physician Michel Allard, who have examined birth, marriage and death certificates of Madame Calment and her family and pronounced them to 'follow each other in perfect order ... with no trace of ambiguity'.

Jeanne Calment was one of a family of four children, two of whom – a sister and a brother – died before she was born. Her second brother lived until 1962 when he died aged 97. Jeanne Calment married in 1896 and was widowed in 1942. She had one child, a daughter, who was born in 1898 and who died in 1934 at the young age of 36. Her only grandchild, a grandson, was also short-lived, dying in 1963 at age 36 like his mother. Since the death of her brother in 1962 and of her grandson in 1963, Madame Calment has had no surviving close family members.

Could she be a fraud? It is hard to see how unless it was the mother and not the daughter who died in 1934, the daughter assuming the identity of her mother. It would have been a cunning trick, for in later years Madame Calment was to reap considerable benefit from her great longevity.

In 1965 Madame Calment made a deal with her lawyer. The lawyer took ownership of her flat in exchange for a monthly pension for life of 2,500 French francs. It is not often that lawyers make mistakes and the deal ought to have worked in his favour. After all, Madame Calment was 90 years old at the time. Little did the unfortunate lawyer know that he would predecease his elderly client and that his descendants, bound by the terms of his agreement, would eventually pay out more than three times the flat's worth to an old lady with a remarkably tenacious grip on life.

But any deception on Madame Calment's part would have required extraordinary prescience and the connivance of surviving relatives and we should banish such thoughts from our minds. No witnesses to her earlier life are available, and no one is likely to exhume her and her relatives for DNA tests to confirm their relationships. She had lost all of her teeth long before her death, so dental records could be no help. Jeanne Calment's word and the extensive documentary detective work by Robine and Allard are the best we can rely on.

Madame Calment's record is remarkable, but it will be broken one day. We know this with the same certainty that we know the current world record for the 1500 metres or the high jump will be broken too. No one suggests that any of the current world records in athletics is the absolute limit of human performance. In exactly the same way, there is no reason to believe that Madame Calment's longevity is the limit of human survival. In fact, although we often talk of a 'maximum life span' for our species, meaning a life span that cannot be exceeded, there is a growing feeling among gerontologists that the concept of a 'maximum life span' does not really make sense, except in the purely empirical sense of the maximum being the longest life span currently known.

Statisticians call such records 'extreme value statistics'. An extreme value statistic is so named because it defines the end of the known distribution. Such a statistic can move in only one direction. Sooner or later, as the size of the sample – in other words, the number of 'attempts' at the record – grows ever larger, the world record for longevity will increase. And it will increase even if the biological determinants of human life span alter not one jot. The only proviso is that the existing record does not butt up against some intrinsic and absolute biological limit, and we will see in later chapters that this is most unlikely.

It is in fact possible, though improbable, that Madame Calment's record of 122 years has been exceeded in the past. The reason it is unlikely is that conditions favourable to extreme old age are comparatively recent and the number of old people is increasing all the time. There are far more centenarians alive today than at any previous age. The number of people who might be candidates for becoming a supercentenarian (110 years plus) was negligible in the past, but is growing all the time. Madame Calment picked a good time to be alive.

But you do not have to go far to find an apparent challenge to Madame Calment's supremacy. A fine example can be found by visitors to Westminster Abbey in the heart of London. In the south transept that is known today as Poet's Corner, and surrounded by graves and memorials of illustrious poets and writers such as Geoffrey Chaucer, William Shakespeare, Charles Dickens, Alfred Lord Tennyson, Samuel Johnson and George Eliot, lies a simple black and white stone tablet on which is inscribed the following:

Tho:Parr of ye County of Sallop. Borne
in A:1483. He Lived in ye Reignes of Ten
Princes Viz. K.Edw.4, K.Ed.5, K.Rich.3
K.Hen.7, K.Hen.8, K.Edw.6, Q.Ma., Q.Eliz.
K.Ja. & K.Charles. Aged 152 Yeares
& Was Buried Here Novemb.15, 1635.

What are we to make of such an extraordinary claim? Who was Thomas Parr, and how did he get here? Could he really have lived 152 years?

Late-nineteenth-century England saw an extraordinary surge of public interest in the question of extreme human longevity. Perhaps the Victorians, triumphing so marvellously in their feats of engineering and exploration, imagined that even the secrets of ageing were within their grasp. Reports of centenarianism appeared frequently in the national and local newspapers, and a lively correspondence on the subject was carried on in the letters page and in the 'Notes & Queries' column of the London *Times*. A book of the period that I have in my possession was originally owned by one Hubert Smith of Bridgnorth. An enthusiast of the subject, Mr Smith pasted to the pages of the book a large number of newspaper cuttings from the 1860s, 1870s and 1880s. The following examples give something of the flavour:

MORE THAN A CENTENARIAN. – The *South of India Observer*, a paper published at Ootacamund, in the Madras Presidency, states that on 16th of March there died, at Ahmednuggur, a venerable Mohamedan at the age of 143 years. He had lived a very religious life, was a priest of his caste, he was never married, and is now naturally considered a saint by the Mahomedans. (*Birmingham Morning News*, 29 April 1874)

UNTITLED. – No person pretending to the possesssion of a well-regulated mind can have any doubt as to the veracity of the *Mensagero* of Mexico; and this fact being conceded, it is accordingly of the highest interest to study the chronicles of that estimable journal. Especially of value is the history it now relates of a venerable gentleman, named Miguel Solis, who lives in the Republic of Salvador, and is actually a hundred and eighty years old. There is no doubt at all about it. He signed a document relating to the building of a convent in the year 1722, being then twenty-three years old. (*Telegraph*, 29 September 1879)

A CENTENARIAN. – Betsy Leatherland, the old lady whose portrait we give on another page, is no less than 111 years old, yet, as we recorded last week, she lately cut a small quantity of corn on the farm of Mr Mead, near Tring, Hertfordshire, in the presence of some hundreds of people. (*Pictorial World*, 15 August 1874)

TO THE EDITOR OF THE TIMES. – Sir, In reference to a letter signed 'A Pilgrim' in *The Times* of this morning, I beg to say that I am assured that there is a woman of the name of Mary Thompson, now, or very lately, living in the parish of Killesher, near Enniskillen in Ireland,

who was born on the 1st of May, in the year 1754. If this is correct, as
I believe it to be, she must now be 110 years old. (*The Times*, 7
January 1865)

Amid all this excitement a sober book with the title *Human
Longevity: Its Facts and Fictions* was published in 1873 by William
Thoms, deputy librarian at the House of Lords. The British House of
Lords, even today, is a good place to study human antiquity because
most of those entitled to hold a seat there do so until they die.
Thoms researched his book, as he describes in his Preface, 'in an
earnest desire to ascertain the Truth, the whole Truth, and nothing
but the Truth, upon this very important physiological and social
question'. His results were startling. Thoms debunked a great many
myths of extreme longevity, including that of Thomas Parr.

What we know for certain of the life of Thomas Parr is this. He
lived in the county of Shropshire near the town of Alderbury. In the
year 1635, the Earl of Arundel, visiting his estates in Shropshire,
received reports of a very old man living thereabouts. He saw Parr,
was impressed by his great age, and sent him, carried on a litter, to
London where in late September he was presented at the court of
King Charles I.

Thomas Parr's life as a celebrity was brief. The smog and stench
of seventeenth-century London quickly destroyed the health of an
old man who had lived his entire life in the country. In just a few
weeks the 'old, old, very old man', as one chronicler described him,
was dead. Parr died on 14 November 1635 and was buried the next
day in Westminster Abbey.

At the order of King Charles, Parr's body was subjected to
autopsy by none other than William Harvey, eminent physician
and discoverer of the circulation of the blood. Harvey's autopsy
report makes interesting reading. Harvey made no claim to verify
the age of the deceased, but confined himself to an examination of
the state of Parr's organs. His lucid account confirms that the body
was that of an old man in a good state of general health, except for

the lungs, which were 'much loaded with blood, as we find them in cases of peripneumony'.

The only point at which Harvey's report commented explicitly on the claims of Parr's extraordinary longevity was when he examined the genitals:

> The organs of generation were healthy, the penis neither retracted nor extenuated, nor the scrotum filled with any serous infiltration, as happens so commonly among the decrepid; the testes, too, were sound and large; so that it seemed not improbable that the common report was true, viz., that he did public penance under a conviction for incontinence, after he had passed his hundredth year; and his wife, whom he married as a widow in his hundred-and-twentieth year, did not deny that he had intercourse with her after the manner of other husbands with their wives, nor until about twelve years back had he ceased to embrace her frequently.

Part of the legend that had grown up around Thomas Parr was that he was a lusty old goat. His conviction for 'incontinence, after he had passed his hundredth year' was not for wetting his pants, but for conceiving a child out of wedlock. For this crime, he was purged by doing public penance, standing wound in a sheet in Alderbury parish church.

Such a report was grist to William Thoms' investigatory mill. Punishments of this nature were set down as a matter of course in the parish records, and it was there that Thoms went in search of Parr's criminal record. It was not to be found. Nor, in spite of exhaustive search and wide advertisement, was Thoms able to find a single shred of reliable evidence to support the claims of Thomas Parr's great longevity. Much indignation was expressed at Thoms' audacity for questioning what so many had, as one correspondent put it, 'on good authority, known to be true', but no one could actually produce the 'good authority' to back up the story.

It turned out that the original version of Old Parr's story was documented only in a verse account, written shortly after his death by one John Taylor, and submitted, elegantly printed and bound, to King Charles. What was the source of Taylor's account? We can be almost sure that Old Parr made it up. William Harvey concluded his autopsy report by remarking that Parr had been blind for the last 20 years of his life, and that 'his memory was greatly impaired, so that he scarcely recollected anything of what happened to him when he was a young man'. All of which goes to show that, when it comes to tall stories about old age, Kings of England can be gullible too.

William Thoms was not, let it be understood, a mere debunker of myths. He sought evidence for, and proved, a good number of genuine claims of centenarianism. In doing so, he established valuable principles for the proper authentication of such claims. Thoms listed the five kinds of evidence he encountered most often: (i) baptismal certificates, (ii) tombstone inscriptions, (iii) the number of the centenarian's descendants, (iv) the recollections of the centenarian; and (v) the evidence of old people still living, who knew the the centenarian as 'very old' when they themselves were quite young.

The last two of these have obvious flaws, and we have seen that Old Parr's tombstone does little to inspire confidence in evidence of type (ii). Tombstones merely record the information given to the stonemason, give or take the occasional slip of the chisel. As for descendants, not only is method (iii) intrinsically unsound, but it is open to all kinds of abuse. Legend credits Old Parr with a son who lived to the age of 113, two grandsons and a granddaughter who lived to 103, 109 and 127, and a great-grandson who lived to 124. But as far as can be found, Thomas Parr left no children, or none that were baptised as his.

This left the indefatigable Thoms with baptismal certificates as the most reliable arbiter of truth, but time and again he demonstrated by painstaking research how easily mistakes could be made. The case of Mary Billinge who died on 20 December 1863, allegedly aged 112 years, is a good example. Her case was investigated at the time of her death, authenticated by the Health Committee of

Liverpool, and duly recorded in the 26th Report of the Registrar-General of Births, Deaths and Marriages. But Thoms discovered that the Mary Billinge who died on 20 December 1863 was not the same Mary, daughter of William and Lidia Billinge, who was born 24 May 1751 and whose baptismal record was accepted as evidence of age at death. She was Mary, daughter of Charles and Margaret Billinge, born 6 November 1772. Mary Billinge was 91, not 112, at the time of her death, which was still a remarkable achievement for someone born in the eighteenth century.

Even with the correct parents, the potential for confusion was not eliminated. Children often died young. The practice of giving the same first name to successive children, by way of perpetuating it, was common. A sad example was the case of Henry and Anne Hibbert of East Yorkshire, who had a son Henry, born 2 July and buried 10 July 1660, another son Henry, born 14 October 1661 and buried 18 August 1665, and yet another son Henry, born 20 January 1672 and buried 16 March 1679. It is understandable if at this point poor Henry and Anne gave up with their unfortunate Henrys and tried luckier names.

When Thoms wrote his book there was a widely held view that centenarianism had been more common during the 'good old days' and that it was also more common among the poor than the rich. Nothing could, of course, be further from the truth. Wealth buys health, and with health goes longevity. Thoms' dismissal of these claims was eloquent and to the point:

> One of the most satisfactory explanations of the popular error, for such it unquestionably is, that Longevity is most frequent among those most exposed to privations and hardships, is that so quaintly described by Fuller in his 'Holy War':
>
> 'Armies both of Europe and Asia (chiefly the latter) are reported far greater than truth. Even as many old men *used to set the clock of their age too fast* [Thoms' italics] when once past seventy, and, growing ten years in a twelvemonth, are presently fourscore; yea, within a year or two after, climb to a hundred.'

This tendency to 'set the clock of their age too fast' is common to old people of all classes alike; but in the higher ranks it is at once corrected by family evidence, the records of the Heralds' College, and similar sources of information, but which leave the self-delusions of the village Hampdens and mute inglorious Miltons unchecked and uncorrected. That in spite of poverty, toil, hard fare and exposure, more in number (not in proportion) of the humbler classes become Centenarians cannot be doubted, but the reason is a very obvious one, namely, that whereas a very limited per-centage of people ever attain to or exceed the age of a hundred, the poor being to the rich as millions to tens of thousands, Centenarianism in the humbler classes preponderates in the same rate over Centenarianism in high places.

Every once in a while a concentration of supposed centenarians is discovered in some far-off place. One of the most renowned has been the Abkhasian people in the Caucasus region of Georgia in the former Soviet Union. In 1972 Alexander Leaf, a physician from Harvard Medical School, returned from a visit to Georgia with photographs and stories of these extraordinary people surviving in good health to great old age, many of them well past 100 years. Zhores Medvedev, an exiled Soviet gerontologist who came to work in London around this time, knew better.

Medvedev revealed that longevity records for the alleged Abkhasian centenarians were non-existent and that few of them were literate, let alone sure of their birthdays. In such a region where old age was well respected, who can wonder that so many set the clock of their age too fast, especially if such a claim found favour with Soviet despot Joseph Stalin, a Georgian himself and a man not noted for his respect of scientific exactitude?

Data on life expectancy around the world are collected on a regular basis by the United Nations in its *Demographic Yearbook* series. These data are compiled from the records of individual member states and vary not only in the frequency with which records are updated, but in the completeness and accuracy of the

Table 4.1 Life expectancies at birth around the world (years)

Country	Men	Women
Algeria	65.8	66.3
Australia	75.0	80.9
Bangladesh	56.9	56.0
Brazil	63.2	70.4
Burkina Faso	45.8	49.0
Canada	73.0	79.8
China	66.7	70.5
Czech Republic	69.5	76.6
Ecuador	67.3	72.5
Finland	72.8	80.2
France	72.9	81.2
Germany	72.8	79.3
Ghana	54.2	57.8
Greece	74.6	80.0
India	57.7	58.1
Indonesia	61.0	64.5
Iran	67.0	68.0
Israel	75.3	79.1
Italy	73.8	80.4
Jamaica	71.4	75.8
Japan	76.6	83.0
Kenya	54.2	57.3
Malawi	43.5	46.8
Mexico	67.8	73.9
Nigeria	48.8	52.0
Pakistan	60.6	62.6
Russian Federation	57.6	71.2
Sierra Leone	37.5	40.6
Sweden	76.1	81.4
Syria	64.4	68.1
United Kingdom	74.2	79.4
United States	72.2	78.8
Venezuela	66.7	72.8
Yemen	49.9	50.4
Zimbabwe	52.4	55.1

SOURCE: *United Nations Demographic Yearbook* (1997).
Note: The entry in this table that most nearly corresponds to
conditions in and around Navrongo, described in Chapter 1, is that for
Burkina Faso, also in the Sahel region of West Africa.

returns. Nevertheless, they provide a global snapshot that is telling (see Table 4.1)

If you want to live a long life, it is as well to be born Japanese and female. The influence of gender on life expectancy has long been known (see Chapter 13) and is reproduced in most countries. For the few countries where this gender difference is absent or reversed, the handful of printed digits in the table betoken widespread suffering and misery for countless women and young girls. It has been recorded that in India today, half a century after the death of Mahatma Gandhi, one of the greatest advocates of human rights that the world has ever known, girls are four times more likely than boys to suffer malnutrition, but fifty times less likely to be taken to a hospital when ill.

We will end this chapter by leaving the realm of human longevity and considering the life spans of animals. The richness and diversity of the animal world is found not only in the glorious variegations of feather, scale and fur, and in the astonishing variety of shape, size and form, but also in the span of time that each species may live (see Table 4.2).

The quality of the data is incredibly uneven, so we must take with a large pinch of salt many of the numbers we see. Even so, there are patterns that jump out at us. Large animals are on the whole longer-lived than small. But birds are long-lived for their size, as are bats. And humans are longer-lived than animals many times their size. Human longevity is but a piece of the bigger jigsaw puzzle of why animals have the life spans they do. If we are to understand human ageing, we need to fit ourselves into this bigger picture.

Table 4.2 *Life spans for various animal species*

Species	Life span (years)
Human (*Homo sapiens*)	122
Tortoise, Galapagos (*Testudo elephantopus*)	106+
Elephant, Indian (*Elephas maximas*)	81
Chimpanzee (*Pan troglodytes*)	59+
Horse (*Equus caballus*)	50
Gorilla (*Gorilla gorilla*)	49
Eagle, Golden (*Aquila chrysaetos*)	48
Bear, Brown (*Ursus arctos*)	47
Gull, Herring (*Larus argentatus*)	44
Monkey, Rhesus (*Macaca mulatta*)	40
Dove, Domestic (*Streptopelia risoria*)	35
Bat, Indian Fruit (*Pteropus giganteus*)	31
Cat, Domestic (*Felis catus*)	28
Pig (*Sus scrofa*)	27
Newt, Japanese (*Cynops pyrrhogaster*)	25
Sheep (*Ovis aries*)	20
Dog, Domestic (*Canis familiaris*)	20
Bat, Vampire (*Desmodus rotundus*)	20
Toad, African Clawed (*Xenopus laevis*)	15
Rat, Black (*Rattus rattus*)	5
Mouse, House (*Mus musculus*)	3

SOURCE:
The figures are based on life span records derived from various sources. For obvious reasons, some of these records are more comprehensive than others.

CHAPTER 5

The unnecessary nature of ageing

Old age is the most unexpected of all things that happen to a man.

Leon Trotsky, *Diary in Exile*

One of the pleasures of working as a gerontologist is that people are usually interested in what you do. I declare this with feeling, having trained first as a mathematician, before switching to a doctorate in biology. Conversations about mathematics tend to be over rather quickly, but gerontology is different. All of us know about – and have views about – ageing. We can all come up with theories to explain it.

Two of the commonest ideas about ageing, however, are wrong. The first of these is that ageing is *inevitable* because we just have to wear out. The second is that ageing is *necessary* and we are programmed to die because otherwise the world would be impossibly crowded. But ageing is neither inevitable nor necessary, at least not in the way that many suppose. This is good news if we hope eventually to improve the condition and quality of life in old age. In this chapter we will examine why these ideas are mistaken in order to prepare the way for looking in the next chapter at why ageing really occurs.

Science sets great store by theories. The key thing about a

scientific theory, as opposed to other kinds of ideas, is that it must be possible in principle to prove it wrong. If a theory is not capable of being falsified by observable facts, as Karl Popper has taught us, it is not part of science. For this reason, there is nothing intrinsically bad about a theory being wrong; indeed, some of the cleverest theories turn out to be wrong. Science is a bit like building a cabin out of logs that seem sometimes not to fit together, or which turn out to be rotten in the middle. The structure must be tested, and tested, and tested again. And if it breaks, you start afresh.

The theories of ageing we have just met are wrong in a particularly potent way because they undermine any edifice built upon them. In the analogy of the log cabin, they affect the very foundations on which the cabin might be built.

The idea that ageing is inevitable forms the basis of the 'wear-and-tear theory', dating from the early years of the twentieth century, which says that ageing has to happen because organisms are complex, and complex things sooner or later break down. In one sense this is just fatalism again, this time in disguise, but advocates of the wear-and-tear theory of ageing cloak their fatalism in a highly respectable principle borrowed from physics, known as the Second Law of Thermodynamics.

I happen to have a great fondness for the Second Law of Thermodynamics because on an everyday level it provides a wonderfully comforting justification for the muddle and mess we make in our lives. It is like the Second Law of Dirty Dishes in the Sink. It is also a sparkling intellectual construct. The Second Law of Thermodynamics is about entropy, or disorder, and it says that in closed systems entropy must increase. We all know about entropy. Entropy is what you get more of if you let slip a stack of neatly ordered papers and they scatter all over the floor, or if, as happened to me one unforgettable day in Mr Hoddle's hospital, you crash the heavy lunch trolley into an unnoticed radiator pipe while manoeuvring past an obstacle on the way from kitchens to ward. In the context of hospital catering, entropy is chocolate sauce on the roast beef and gravy on the ice cream. Entropy does not go down well with a ward full of hungry patients.

The wear-and-tear theory of ageing holds that, in living organisms, entropy is necessarily on the increase too. It is easy enough to empathise with this, but it is fundamentally wrong. It is wrong precisely because it fails to pay close enough attention to the Second Law of Thermodynamics. The Second Law of Thermodynamics, remember, is about closed systems. A closed system is like the inside of a sealed chamber. Nothing goes in and nothing comes out. But organisms like you and me are not closed. We are open. To put it bluntly, we have a hole at each end. We take things in and we pass things out, and in making this trade we continually steal from our environment. What we steal is energy, and energy can be used to combat entropy.

There is absolutely no reason why a living being must bow to the Second Law of Thermodynamics, in the sense that it has to grow old and die simply because it cannot manage to keep going. This is not just a theoretical concept. We saw in Chapter 3 that sea anemones and hydra appear not to age. If we accept that a hydra does not age, then it cannot be falling victim to the Second Law of Thermodynamics.

There is another reason to believe that organisms can escape the Second Law of Thermodynamics: if this were not the case, none of us would be here today. That goes not only for humans, but for all other life forms as well. Some billions of years ago, the first cells evolved. Cells reproduce by dividing, and all of the material that forms the new cells comes initially from their parents. Every cell alive today could, if the records existed, trace its ancestral lineage back to those very first cells. At every link in this very long chain, there was continuity of the cellular ingredients.

What this means, and personally I find the thought quite awe inspiring, is that each cell in my body (and in yours too) is a product of an unbroken chain of cell divisions that extends backwards in time to the very beginnings of life on earth. The cellular ingredients, which have been renewed and replaced many times over, cannot have been accumulating progressive wear and tear or they would have died out long ago. This proves beyond the slightest shadow of doubt, as far as I am concerned, that the Second Law of Thermodynamics does not automatically condemn living systems to death by wear and tear.

What applies backwards in time, however, does not apply forwards. Each one of my cells shares this noble ancestry, but most of them are going nowhere. When I die, my cells will die too. The only cells to continue the lineage are the sperm that helped form my children.

The remarkable chain of cells that connects us to the past and that may, with luck, connect us to the future is known as the germ-line. In a human being, the germ-line is found in the reproductive cells of the gonads. In a man these are the sperm and sperm-forming cells in the testes, and in a woman they are the ova, or eggs, and the cells in the ovaries that form them.

The germ-line is immortal. This is not to say that individual germ cells cannot die. Many do die, including the millions of sperm that fail to fertilise an egg. In some respects germ cells even age, which is why genetic abnormalities are commoner in the offspring of older parents. But aside from the increased risk of genetic abnormalities, the child of old parents is the same as the child of young parents.[3] However old its parents, each newborn child begins life with its age clock set to zero. To permit this to happen, it is essential that somewhere within the germ cells there are mechanisms to prevent ageing from occurring.

Most of the cells in the body do not belong to the germ-line. They are the cells of the soma, which is everything apart from the gonads. Somatic cells take the brunt of the ageing process, but in the long term this does not matter because, unlike germ cells, somatic cells are confined to the lifetime of the individual. Somatic cells contribute nothing directly to future generations, or, at least, this was the case until Dolly the sheep was cloned. A large part of the scientific excitement about Dolly was that somatic cells are effectively sterile. Nevertheless, somatic cells have basically the same DNA, proteins and other cellular ingredients as germ cells. What the scientists at Edinburgh's Roslin Institute showed was that injecting the nucleus (the DNA-containing part) from a somatic cell – a mammary gland cell – into an egg from which the nucleus had been removed, resulted in the 'fertilisation' of the egg and the birth of an apparently normal lamb. Dolly serves graphically to underline the central puzzle of gerontology, which is how germ cells manage to keep themselves immortal while somatic cells undergo the ageing process. What Dolly has shown is that there is something about germ cells that can apparently override the ageing of somatic cells. The idea that wear and tear is an inevitable feature of living systems is not a good enough explanation of ageing.

'Wait a minute,' you might say. 'What about the brain? The brain has an order and complexity that must run down in time. Brain cells don't divide, but they do die. Sooner or later this must kill you.'

This is true, but only if we accept that the present-day construction of our brains is the way brains have to be. What this observation fails to do is explain why the brain does not replace its dead cells, as other organs do, or make better running repairs to its cells so that they last longer.

The real trouble with the wear-and-tear theory is that it offers an explanation of *how* ageing is caused, but it fails as an explanation of *why* ageing occurs. Why do some species age and others do not? Why is the germ-line immortal and soma not? Why do different species, subject to the same general kinds of wear and tear, have such different life spans?

'Why?' questions are the province of evolutionary biology and the second wrong idea, that ageing is necessary to prevent overcrowding, takes us into this realm. The idea forms the basis of what we will call the 'death gene theory', which suggests that ageing has evolved through the action of Darwinian natural selection because in some sense ageing is a good thing.

Charles Darwin's theory of evolution by natural selection has immense explanatory power, but like any powerful instrument it can be damaging if misapplied, so we will review it first, before we try to use it. It is as well to appreciate that natural selection is really three things rolled into one.

In its abstract form, natural selection is a mathematical law. If we postulate the idea of genes that replicate themselves, and if we suppose that genes vary in their effects on survival or reproductive success, then natural selection simply has to happen. Those genes that confer the highest survival or reproductive rates will tend inevitably to predominate. Put like this, the theory is even a tautology and has been criticised as such: if natural selection is true by definition, then it cannot be tested and so does not count as scientific theory.

But the theory *can* be tested because natural selection is also a hypothetical process that we ought to be able to detect in the world around us. We can set up situations where we artificially impose a selection for some particular attribute, as when we breed dairy cows for increased milk yields or potatoes for resistance to blight. These kinds of experiment will not work if the theory is wrong.

A good example of the *process* of natural selection occurred in the north of England following the industrial revolution, when not only buildings but also trees became darkened with soot. Moths that previously matched the original tree colouring very well found themselves standing out against the new, darker backgrounds and being picked off by predators. Because there was genetic variation in the darkness of the moths' wings, the darker varieties did better and selection ensured that in time the entire population of moths evolved darker colouring. Since the Clean Air Act and the closure of many industrial mills in the 1960s and 1970s, the trees are getting lighter again, and the moths are following suit. The full process of natural selection is vastly more complex than this, and we still understand it only partially.

The famous phrase 'survival of the fittest' – coined by Herbert Spencer and adopted by Darwin – has a neat ring to it, but it is not entirely specific. By 'fitness' do we mean the ability to flood the environment with offspring, or to avoid extinction during ecological catastrophe, or what? And whose fitness do we need to consider: the species, the individual or the gene? Scientific understanding of the process of natural selection is becoming ever richer as we gain deeper knowledge of how genetic information undergoes change and reassortment as it is used and passed on to the next generation.

Lastly, natural selection is an *explanation* of how evolution has occurred. We can look at the wing bones of a bat and suggest how the forelimb of a mouse or rat has adapted to provide the means of flight. We can understand how the bat's ancestors might have gained advantage by the ability to evade ground-dwelling predators. We can look at a flower and a bee and suggest how each has adapted to the presence in its environment of the other, to the mutual benefit of both. We can look at different species, note their similarities and differences, and sketch family trees of how they might have diverged from a common ancestral stock. We can dig in the ground and find the fossilized remains of species now extinct which might have been these ancestors.

It is as an explanation of the history of life that natural selection meets its greatest opposition, chiefly from those who find it contrary to their religious beliefs. Once, on a train ride from Los Angeles to Santa Barbara, I shared a compartment with an 11-year-old girl and her parents. The girl was bright and outgoing and she chattered away to me, while her parents gazed out at the passing scenery. But when she asked me what I did for a living and I answered that I was a scientist interested in why we grow old, her face darkened in anger.

'Oh,' she said. 'You are one of those stupid people who waste their lives trying to discover what God knows already,' and she spoke to me no more.

The truth is that biology makes sense only if one accepts that life evolves by natural selection. If you believe in a Creator, then the evidence is strong that the biosphere was created to evolve by natural selection. Either way, the theory of natural selection is one of the best tools we have to understand the living world.

Ageing, some theorists have suggested, takes place because we have evolved death genes (yet to be discovered) to destroy us when our time is up. The essence of the argument is that, if we did not have such genes, the population would grow without limit, exhaust its food supply and experience general starvation. Like lemmings over a cliff, we throw ourselves into the grave for the good of the species. This is an idea that no doubt would have appealed to Dr Pangloss, the eternal optimist in Voltaire's splendid satire *Candide*, who declared that everything is for the best in the best of all possible worlds. Pangloss believed, for example, that the human nose has the shape it does because it is made for spectacles to sit on. It does not find favour with evolutionary biologists.

If we examine more closely this idea that ageing has evolved as a form of population control, we will discover how and why it is wrong. The theory is often found to hinge on the assertion that ageing is necessary to rid the population of old and worn-out individuals who would otherwise compete for resources with their younger and fitter progeny.

A moment's consideration shows that this is a circular argument. It assumes that the old are decrepit, when this is the very state of affairs we are trying to explain. We can try to correct the more obvious fault by modifying the assertion as follows: 'Those who are past reproductive age are a useless burden to the species. What can be more natural than that nature does away with them?' But the argument is still circular because, while it is certainly true that old animals eventually lose their powers of reproduction, this surely is the consequence of ageing, and not its cause (we will consider reproductive ageing more closely in Chapter 11).

And so we might come to the following statement of the death theory:

Ageing is a means for a species to regulate its numbers, so that it can thereby avoid the overcrowding of its environment, which might otherwise lead to its demise. In fact, one could go further and suggest that ageing facilitates the evolution of other benefits. For natural selection acts upon the variations among progeny at each generation, and change can take place only when generation succeeds generation. Therefore, to the extent that ageing limits the span of each generation, it aids the adaptation of species to their environment and is a good thing. Because ageing is beneficial in this way, we may suppose that there have evolved genes whose function it is to draw life to a close, and that it is the action of these genes we observe as senescence.

This theory is about as wrong as any evolutionary theory can be. Firstly, there is almost no evidence that ageing actually does what the theory suggests: namely, act as a mechanism of population control. The reason is simple. Animals in the wild do not as a rule live long enough to grow old. The wild environment is a dangerous one and most animals die young from hazards like predators, starvation, accidents and disease. For nearly all species, the problem is how to keep population sizes up, not down. A few of the larger mammals, such as elephants, and some long-lived birds show some evidence of ageing in the wild. But it is hard to be sure whether the present-day conditions of elephants in game reserves really represent the environment of their evolutionary past. In any case, it does not matter that a handful of species show some age-related mortality. The fact that most do not, and that even for those which do, ageing makes only a minor contribution, is sufficient to make it highly implausible that the reason for ageing is that it is necessary to control population size.

A second objection to the death gene theory is equally devastating. The theory is based on the notion of natural selection working for the 'good of the species' when this runs counter to the interests of the individual. Selection of this kind simply will not work. We can see the problem clearly if we conduct a little thought experiment, an imaginary exercise to examine how a population would evolve in the presence of a death gene that causes ageing.

Imagine a population in which the death gene for ageing has evolved – say, to protect the population from overcrowding (we ignore for the moment that there is scant evidence that this actually happens). We suppose that ageing is universal among the population, so that all the individuals possess the death gene; they dutifully lay down their lives for the good of the population at the appointed time. Now remember that the copying of genes is not 100 per cent accurate. From time to time there will arise a mutation that inactivates the death gene. Any individual carrying the inactive death gene will on average live longer than its companions, and since it does not age, it will on average produce more offspring. These offspring will themselves carry the inactive death gene, so the mutant gene will steadily increase in frequency within the population. The result will be that freedom from ageing will spread, favoured by natural selection.

In the early stages of our thought experiment, most of the population still carries the intact form of the death gene, so if it does confer any benefit for the population as a whole, this will also benefit the minority of immortal mutants. In effect they are freeloaders, or cheats. As time goes by and the mutants come to predominate, we might ask whether the removal of the restraint on population growth, which ageing previously provided, will become a problem.

The answer is that, even if the population does begin to suffer overcrowding, this is not going to restore the death gene. Ageing and non-ageing individuals are assumed to be similar in all respects other than the presence and absence of ageing, and they suffer equally from the effects of overcrowding. All that can happen is that, as the population grows in size and the death rate from starvation rises, due to pressure on finite sources of food, this will diminish the advantage in being immortal. But it will not be enough to turn the tables and bring the death gene back, because there is no selection advantage to the *individual* in being mortal.

In short, the death gene theory does not fit the facts. It predicts that ageing should be an important contributor to mortality in the wild, which it is not, and it depends upon an unworkable model of natural selection. We must set it aside.

This conclusion has sweeping implications. The last few decades have seen great progress in an area of research known as developmental biology. Developmental biology is concerned with the genes that control the unfolding of the organism from fertilised egg to fully functioning adult. The genes that do this exhibit exquisite timing and co-ordination. Many have sought to interpret ageing as an extension of the developmental process and suggested that genes of a similar kind are responsible for the timing and co-ordination of our ageing and death. But if ageing is not controlled by death genes, this cannot be right.

We arrive at a conundrum: ageing is neither inevitable nor necessary. So why does it happen?

CHAPTER 6

Why ageing occurs

If I'd known I was gonna live this long, I'd have taken better care of myself.

Eubie Blake, on reaching the age of 100

Under a railway arch in the grounds of the University of Manchester Institute of Science and Technology is a statue of Archimedes leaping out of his bath. The expression of surprise and delight on Archimedes' face is so real that you can almost hear a cry of 'Eureka!' echo from the brickwork. Archimedes, you will recall, had just worked out his famous Principle, which explains, among other things, why a boat floats in water. In homage to Archimedes, I have always found the bath a good place to think and it was there, one February night in 1977, that it suddenly dawned on me why ageing occurs.

The reason that you and I will grow old and die, I am sorry to say, is that we are disposable. And the saddest thing is that this assessment of our disposability is made by none other than our very own genes.

It is not that our genes actively destroy us – as we saw in Chapter 5, this does not make biological sense. It is that the interest that our genes have in keeping us going does not happen to coincide exactly with our own. This, then, is the bad news. The good news is that, once we realise what it is that our genes are up to, we are on track to discovering what it is that actually causes us to age. As we shall see, instead of being deterministically time-limited by some fateful hourglass, there is a plasticity to the course of our lives that we may be able to turn to our advantage.

At the time of my bath, I had been puzzling over the problem of why normal human cells grown in the test tube invariably age and die, and about what this might have to do with the ageing process as a whole (see Chapter 7). I had also been grappling with the idea that cells might undergo this ageing process because they were vulnerable to a hypothetical process known as 'error catastrophe'. Error catastrophe is a bit like the dreadful shriek you sometimes get in a public address system when the microphone picks up noise from the loudspeakers, and the sound gets amplified around and around the loop. In cells, feedback loops exist in the molecular machinery that makes DNA and proteins, because the same machinery also makes new copies of itself. A mistake at one step can create a defective unit, which in turn makes even more mistakes, and so on and on. The possibility therefore exists that mistakes can be amplified around these loops. Some mathematics had shown that error catastrophe was a theoretical possibility, but that cells could avoid it.

The question I was pondering among the soap bubbles was why cells might succumb to error catastrophe if, as my calculations showed, they did not need to. 'Why?' questions were much on my mind at the time, for I had recently met John Maynard Smith, grand master of evolution theory, and been busy reading a series of his papers with provocative titles like 'What use is sex?' and 'Why be a hermaphrodite?'*

Two other concepts were also revolving around in my head. One was an idea proposed two years earlier by scientists in California and in Paris to explain how cells manage to make proteins as accurately as they do. Protein synthesis is like stringing beads, but the trick is to select the right bead each time from the twenty

---

* Hermaphrodite species like snails have both male and female sex organs in the same individual.

different kinds that are available, and to do this at speed using rather basic molecular machinery. The new idea, called 'kinetic proofreading', showed that cells could in principle be as accurate as they liked, but only at a cost. The cost was the use of extra chemical energy that would be needed to fuel the high-accuracy selection process. This was important because the model of error catastrophe showed that cellular meltdown could be avoided if proteins were made very accurately.

The other concept in my mind was the much older idea of the germ-line and soma distinction, which was introduced into biological thinking in the late nineteenth century by the German naturalist August Weismann. As we saw in Chapter 5, the germ-line must be immortal if life is to continue, but no such consideration applies to the soma.

What I realised in my bath was this: a multicellular organism needs a lot of accuracy in its germ-line, which must transmit its genes to the next generation, but it does not need so much accuracy in its soma. Sooner or later the soma is going to die by accident. Might it not be better to save energy and make somatic cells in a more economical way, even if this results in them ageing? The answer to this question hinges on how long you need the soma to last.

Let me illustrate this with a simple analogy in the form of a fairy tale. Once upon a time in the far-off land of Senescia, there lived a princess who wished to choose a husband. Three eligible princes gladly offered themselves as candidates and the princess, finding them all equally attractive, chose to set them a competition. Each of the princes was given a hammer, a chisel and a very large pile of stones. The winner would be the one who in 24 hours could build the greatest number of stone towers 1 metre square and 2 metres tall.

Prince Albert, eager and rash, cast his hammer and chisel aside and piled the stones one on top of another just as fast as he could manage. But his towers wobbled and fell, and when the 24 hours were up, poor Albert had built six towers, but all he had left were six jumbled heaps of rock.

Prince Brian, a methodical fellow, set to work at once with hammer and chisel to shape each stone into a fine rectangular block. When the time was up, Brian had a marvellous structure that would surely last a lifetime, but only half of a single tower was complete.

Prince Cedric, a canny chap, used his hammer and chisel to knock a few stones quickly into shape for the base and corners of each tower, but he filled in the walls between with rough, unhewn stones. After 24 hours, Cedric had two towers standing, albeit with slight cracks beginning to show, and a third tower half built. Cedric was the undisputed winner of the hand in marriage of the prudent Princess of Senescia. And they lived happily ... (as we saw on p. 12, you should choose your own ending).

The point of this tale is that Cedric's corner stones correspond to the germ-line, whereas the rough infill of his walls corresponds to the soma. His strategy was the winning one because it optimised the deployment of his time and energy differentially between the two. It was an idea along these lines (minus the fairy tale) that occurred to me as I bathed.

Eureka! moments in science are few and far between, but when they come they repay the hard grind that precedes them. I got out of my bath, dressed and sketched the bones of my idea on paper. Early the next day I was to leave for Sweden where I would be working on a quite different problem, and it was several days before I could give my full attention again to this idea about ageing. When I did, I wrote a detailed scientific account and discussed it with Robin Holliday, a geneticist who introduced me to the problem of ageing, and with whom I collaborated for several years. Holliday liked the paper, but he declined my invitation to co-author it. I then sent it to John Maynard Smith and with some trepidation waited for his comments. To my relief, Maynard Smith approved the paper in terms that made me confident I was on the right track, and I submitted it to *Nature*, the premier international journal of science.

No scientific paper in a quality journal gets published without peer review by expert referees and that process takes a while. I finally nerved myself to ring the *Nature* editorial office and ask whether a decision was expected soon. The biological sciences editor came on the line. 'Oh yes,' she said. 'The last referee's report just came in. We'd like you to shorten the introduction and make a couple of other small changes, then we'll accept it.' I was thrilled.

The article was published on 24 November 1977 with the title 'Evolution of ageing'. At the time, I was working neither as a gerontologist nor as an evolutionary biologist, but in the National Institute for Biological Standards and Control in London, concentrating mainly on the measurement of proteins in the blood that make it clot, such as Factor VIII, the enzyme that is lacking in haemophiliacs. The work was interesting and, it being my first job, I was lucky to have a director who foresaw that my enthusiasm for ageing would spill over into the other areas of my work, as it did. But this double life had its quirks. Two weeks after my paper came out in *Nature*, I was at a conference on haemophilia when the conversation at dinner turned to ageing. One of my table companions asked if I had read the recent article in *Nature* on the evolution of ageing, and before I could answer, he gave his opinion of it. Luckily for both of us, this was favourable!

Reaction to the theory filtered through to me from the gerontological community, and more and more I found myself invited to conferences to speak about ageing. I have to say that my ideas did not find immediate favour in this camp. In the late 1970s, most gerontologists inclined to the view that ageing was programmed, like development, and only a minority supported the view that ageing was 'stochastic' – that is, driven by the chance accumulation of mistakes. My theory offered evolutionary support to the stochastic view and because of this it was met with some scepticism. But at least people listened, and gradually the idea that the programme theory did not make evolutionary sense began to take hold.

The theory needed a name and it seemed natural to call it the 'disposable soma' theory. The name was prompted, of course, by comparison with disposable products like coats, cars and washing machines. The manufacturer of such products has to think carefully about how much to invest in their long-term durability. If the product is shoddily built, it will fall apart rather quickly, like Prince Albert's towers, and the customer will be dissatisfied. On the other hand, long-term durability increases the manufacturing costs and makes the product less competitive in the market place, as Prince Brian found to his cost. Successful manufacturers, like Prince Cedric, reach the right compromise: the product lasts long enough in good condition to satisfy the customer, but not for ever. So it is with our bodies, at least from our genes' point of view.

I soon realised that the disposable soma theory had much more to say about ageing than just to support error catastrophe. Error feedback in cells is only one way that organisms might age. I published further work showing how the idea would generalise to cover the whole spectrum of maintenance and repair activities that an organism might need to carry out on its soma. Maintenance is good for the organism because it aids survival, but it is also bad because maintenance activities, like everything the body does, require energy to fuel them. In fact, the disposable soma theory exemplifies the idea of a trade-off in the 'bioeconomics' of living systems.

In bioeconomic terms, we can think of an organism as an entity that takes resources from its environment – primarily energy in the form of nutrients – and produces a genetic output in the form of its progeny. Darwin's basic concept of natural selection – survival of the fittest – can then be rephrased as the statement that those organisms which are most efficient at converting resources into viable progeny, under the constraints of the ecological niche in which they live, will become the most numerous. In genetic terms, the genes specifying the most effective use of resources will come to dominate the gene pool.

The bioeconomic problem faced by an organism is just like the difficulty that you and I face when we try to work out how best to use our money: it is a problem of allocation. Just as money spent on a holiday in Florida cannot be used to repair the roof, so energy spent on producing progeny cannot be used to repair DNA in the cells of the soma.

To appreciate the power of the bioeconomic analysis to explain why we age, let us conduct a new thought experiment. This time we are going to begin with a hypothetical population of immortal animals. We will call our imaginary animals 'mobbits', and should think of them as a bit like a mouse and a bit like a rabbit.

Mobbits are immortal only in the sense that they do not age. They can die of accidents, such as starvation, cold, infection or being eaten by a predator. Let us suppose that each mobbit has a 50 per cent chance of making it through the next year of life, and a 50 per cent chance of dying.

Because mobbits do not age, a 1-year-old mobbit is physiologically identical to a 5-year-old mobbit, and a 5-year-old is identical to a 20-year-old. We cannot tell them apart. But if we tag each mobbit with a label indicating when it was born, we will notice very quickly that there is one major difference between 1-year-old, 5-year-old and 20-year-old mobbits: 20-year-old mobbits are very much rarer. While half of all mobbits will make it to their first birthday, only one in thirty-two will make it to 5 years of age, and only one in a million will make it to age 20!

Does this difference matter? Yes, it matters a lot, and it matters because of the resources invested in maintenance of the soma. To make a mobbit immortal takes lots of maintenance. But why should each mobbit invest so much in maintenance when only one in a million of them will still be around to celebrate its twentieth birthday?

Suppose a mutation happens in a mobbit's DNA that reduces its investment in somatic maintenance and makes more energy available for growth and reproduction. Let us say that the mutant mobbit is no longer immortal, but ages and dies when 20 years have passed because of an accumulation of somatic defects. Only one in a million of the mutant mobbits will know the difference in terms of its impact on survivorship, but all of the mutant mobbits will enjoy increased rates of growth and reproduction! It is clear that natural selection will favour the mutant. The mutation will spread, and ageing will enter the population.

Let us compare this scenario with the thought experiment of the previous chapter, where we showed that, if we assumed ageing was programmed by a death gene for the good of the species, then we would expect immortal mutants to multiply and take over the population. In the present case we have the reverse situation: *mortal mutants spread through an immortal population*. The crucial difference lies in the way natural selection is seen to work. Previously, we began by supposing that ageing was good for the species in order to explain how something that is patently bad for the individual might have evolved. We found that this could not work.

In the present case, we make no prior assumption about whether ageing is good or bad. Of course it is bad, but it comes about as the result of a trade-off. Long-term survival is sacrificed by the genes in favour of the biological imperative of reproduction. In the wild environment, this does not matter terribly much because the organism is likely to die before ageing takes effect, so the fact that ageing has entered the life cycle passes almost unnoticed. But if we now capture some of the mobbits from the wild and keep them as pets, protecting them from the dangers of the wild, we will find that they exhibit ageing because their somatic maintenance systems are no longer sufficient to live indefinitely (see Figure 6.1).

A similar argument explains why humans are disposable too. Our somatic maintenance systems evolved at a time when the conditions of life were much harsher, and few could expect to survive to old age. The catch from the human point of view is that, in the space of only a few thousand years, we have developed all

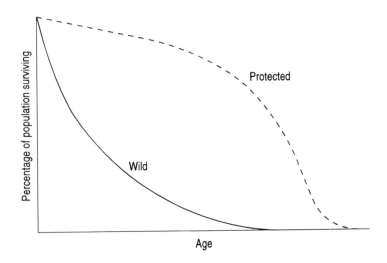

FIGURE 6.1
The contrast between the survival curves for populations of animals under protected conditions and in the wild. The effects of juvenile mortality, which are not included here, would make the difference even more striking.

sorts of advances that have reduced our level of accidental mortality rather dramatically. We have tamed our wild environment and continue to do so. The price we pay for the fact that our social and cultural evolution has outstripped the pace of our biological evolution is that we experience ageing to an extent that has probably never occurred in the living world before. We are, as American physician Boyd Eaton has put it, 'stone-agers in the fast lane'.

The disposable soma theory tells us convincingly why we and other animals age, provided we make just two assumptions. First, the level of accidental mortality must be greater than zero, which is surely reasonable to suppose. Secondly, we require a distinction between germ-line and soma. Remember the hydra. The reason why the hydra is immortal is that its germ-line permeates its soma, so that the two are effectively one. Abraham Trembley showed this in the eighteenth century when he found that he could cut up the hydra and it would regenerate a whole animal from even a small part. In effect, it has no soma to be disposable. Any animal that can reproduce in this vegetative manner is likely to be similarly immortal, as are many plants.[4]

The disposable soma theory can do much more than explain why ageing happens. It can guide us in the study of the ageing process itself (as we shall see repeatedly in later chapters of this book). Above all, the theory tells us that ageing is probably due to the gradual and progressive accumulation of damage in the cells and tissues of our bodies as we live our lives. It tells us that ageing is unlikely to be caused by only a single mechanism, because it applies equally to all the different maintenance and repair systems of the soma, suggesting that imperfections in each of them will play a part. And it tells us something very important about the way in which genes regulate duration of life.

In the remainder of this chapter, we will briefly extend our thought experiment with the mobbits to examine why different species have the life spans that they do. Our thought experiment is, of course, hypothetical, so we will also track back in evolutionary time to consider the true origins of somatic mortality. Lastly, we will fit the disposable soma theory into the context of some earlier ideas on the evolution of ageing. But before doing this, I want to lay to rest an erroneous objection to the disposable soma theory which crops up from time to time. The objection is this: Is maintenance really so costly in energy terms? Are resources really so scarce that organisms cannot invest enough in maintence to keep going indefinitely, as well as reproducing?[5]

The answer to the first question is a resounding yes. Maintenance in its totality is expensive. It is only quite recently that biochemists have discovered just how much energy gets spent on basic tasks like making proteins accurately and clearing away the debris that accumulates in cells every day of our lives. But we can also gauge how much maintenance costs in another way. Each of us takes in a sizeable intake of energy each day. We burn only a little of it in physical work. We do not grow (or at least not very much). So where does that energy go? It goes for all of the everyday metabolic processes that we tend to take for granted, a lot of which are concerned with maintenance of one kind or another.

To answer the second question, think again of the mobbits. But now suppose that the mobbits are surrounded by more food than they can eat: lettuce, carrots and grains at every turn. Energy is abundant, but even so the mobbits have only a finite influx of energy at any given time. They still have a problem of allocation. If accidents still happen, and they will, it is never going to be worthwhile to be immortal.

To explain why different species have the life spans that they do, we first need to identify what life span we might expect for our hypothetical mobbits. We need to resume our thought experiment and run it just a little bit longer.

We saw how the 20-year mutant mobbit would be fitter in evolutionary terms than the immortal mobbit, but the 20-year mobbit is still spending too much energy on maintenance. One in a million is a tiny number, so why not reduce the investment in maintenance still further? Suppose a new mutation gives the mobbit a 10-year life span and a further increase in reproduction rate. The chance of surviving 10 years is still tiny, only one in a thousand. The new mutation results in an even faster spread of its genes and, once again, the strategy is a winner. The 10-year mobbit multiplies faster than the 20-year mobbit and takes over the population. And so it goes on, but not for ever.

Eventually there comes a point where any further reduction in maintenance will begin to hurt. Remember there is a one in thirty-two chance of the mobbit still being alive at its fifth birthday, which is no longer so negligible. Natural selection will eventually reach an optimum balance at the point where any further gain in reproduction is cancelled by the growing loss in terms of diminished survival. For the mobbit with its 50 per cent accidental mortality a year, this optimum might be at a level of investment in maintenance that gives a 6-year life span.

With this background, we can now readily see how animals have evolved different life spans. Suppose a mobbit mutated its form to acquire the means of flight, as bats evolved from ground-dwelling rodents. Or suppose the mobbit evolved a protective shell like a tortoise or turtle. Suppose too that the modification brought reduced accidental mortality, say from 50 per cent per year to 30 per cent. The chance of a modified mobbit escaping accidents over a 10-year time span then jumps from a negligible one in a thousand to a rather more significant one in thirty-five. Because of this, it no longer makes sense to age at 6 years and it becomes worthwhile to push the investment in maintenance up a little, so as not to waste the benefit of the modification by ageing too fast.

The outcome is that organisms exposed to high risk are predicted to invest rather little in maintenance and a lot in reproduction, whereas organisms exposed to low risk are predicted to do the opposite. The low-risk organisms become the Rolls-Royces of the animal world, the high-risk organisms become the————'s (insert your own low-cost make of automobile here). There is good evidence that this is so.

Bats live longer than rats, but reproduce more slowly. Birds on the whole live longer than ground-dwelling mammals, but flightless birds have short lives. And some of the larger species of tortoises and turtles in their armoured carapaces live longer even than humans.

The disposable soma theory predicts that, if we look at maintenance systems in long-lived animals, we will find them set higher than maintenance systems in short-lived animals, and such indeed seems to be the case. Some years ago, American biologists Ronald Hart and Richard Setlow showed that cells from long-lived mammalian species repair their DNA more efficiently than those from short-lived species. In my own laboratory, we have recently extended this observation. Using cells grown from skin biopsies from a number of different species, we have found that cells from long-lived animals are better protected against challenge from a range of chemical and other stresses than cells from short-lived species.

The mobbits tell us why ageing occurs and why species have the life spans they do, but we should not suppose that immortal mobbits (or animals like them) have ever existed. If we want to find the real evolutionary origins of ageing, we must look further back into the remote past when multicellular life first hit upon the strategy of division of labour, and designated some cells to take on the reproductive functions of the germ-line and other cells to take on the specialised tasks of the soma. A nice illustration of this primordial division of labour can still be found in present-day microscopic creatures called Volvocales.

The Volvocales are flagellated forms of green algae, comprising a variety of species whose Latin names mostly begin *Volvox*. Flagellated in this context means possessing whip-like flagella (singular: flagellum), which the Volvocales thrash about to drive themselves through water.

Green algae are denser than water, but need to stay near the surface to get light for photosynthesis, so the flagella are essential to keep afloat. But herein lies a problem. A cell would find it extremely awkward to keep beating its flagellum, anchored as it is to a molecular motor in the cell wall, while getting on with the business of cell division. This would be a bit like trying to change your clothes at the same time as playing a hard game of tennis. Life is tough. Do you ditch your flagellum and risk sinking into oblivion, or do you hang in there, beating your flagellum and wondering if you will ever get around to reproducing?

The Volvocales solved this problem by becoming multicellular and evolving a cellular division of labour. Germ cells sit in the middle and divide. Somatic cells beat their flagella and keep the whole show afloat. When the embryos formed from the germ cells are released, the disposable soma is abandoned and dies. It was perhaps in a scenario like this that ageing first appeared on this planet.

It was an idea along these lines that August Weismann began to explore back in the 1880s, and which resurfaced in gloriously bizarre form in the writings of Elie Metchnikoff some 20 years later. Metchnikoff was a Russian scientist who worked much of his life at the distinguished Pasteur Institute in Paris, and who in 1908 won the Nobel prize in physiology and medicine for his work in immunology. Metchnikoff became deeply interested in ageing and suggested that a primary cause of senescent decline was the action of toxins secreted by the bacteria in the gut.

Incidentally, the yoghurt industry owes a great debt to Metchnikoff. At the time, a myth circulated that great longevity was to be found among Bulgarian peasants who consumed large quantities of a form of soured milk, called 'yahourth'. Metchnikoff succeeded in obtaining a sample of the fabled 'Bulgarian bacillus' used in the production of yahourth and he declared it 'the most useful of microbes which can be acclimatised in the digestive tube for the purpose of arresting putrefactions and pernicious fermentations'. Such was the influence of Metchnikoff's ideas that yoghurt, as yahourth came to be called, soon became a major health fad in western Europe, and has remained firmly linked to the notion of good health ever since. Metchnikoff's acceptance of Bulgarian claims to centenarianism was uncritical, however, and it is clear from his published work that he believed many such claims, including that of the notorious Thomas Parr.

In keeping with his ideas about the toxic effects of bacteria in the gut, Metchnikoff saw the large intestine, or colon, in mammals as the source of life-shortening evil. To explain its existence, he suggested that the organ would 'simply have the function of a reservoir of waste material', in order that the animal would not have to defaecate too often. The reason why mammals are generally shorter-lived than birds, Metchnikoff proposed, could be found in anatomical and physiological differences in their digestive tracts, mammals being handicapped thus:

In order to void the contents of the intestines, mammals have to stand still and assume some particular position. Each act of this kind is a definite risk in the struggle for existence. A carnivorous animal which, in the process of hunting its prey, had to stop from time to time, would be inferior to one which could pursue its course without pausing. So, also, a herbivorous animal, escaping from an enemy by flight, would have the better chance of surviving the less it was necessary to stand still.

Hence the harmful colon. Pigeons, of course, do it on the wing, as many of us will have learned at one time or other to our considerable chagrin.

The mental image conjured up by Metchnikoff's passage above, in which lion and wildebeest pursue each other desperately from toilet stop to toilet stop across the plains of the Serengeti, is delightfully ridiculous. But it makes perfect sense to suppose that there may exist trade-offs that sacrifice longevity for other traits which will increase an organism's evolutionary fitness.

The brain, as considered briefly in Chapter 5, is a good example. The evolution of a brain has obvious benefits. But a brain works best by developing fixed networks of connections between the neurones. This makes the repair of structural damage rather difficult, if not impossible. The longevity of the human brain, in which individual neurones may work for 100 years and more, is no doubt due to rather special mechanisms of cell protection and repair. Good as these mechanisms are, they do not include the replacement of cells that are actually destroyed.

The limited repair capacity of the brain is but one example of a very general idea put forward by George Williams, evolutionary biologist at the State University of New York in Stony Brook, in 1957. Williams pointed out how genes that have good effects in young animals will be favoured even if the same genes do harm later. The example considered by Williams in his paper was of a hypothetical gene that helps calcium to be deposited in the bones of a growing young animal, but which in the adult leads to a gradual hardening of the arteries.

What Williams' idea shows is how the concept of trade-offs can include all sorts of design constraints that might in time affect the ageing organism. Coming from a different area of biology, I had not encountered Williams' work when I had my own ideas on why ageing occurs back in 1977, but when I did read his paper, I could see how well our concepts fitted together. He came at the problem from thinking about the kinds of gene whose effects might eventually harm the ageing organism, and so shape the ageing process. I came at it from thinking about the mechanisms that keep an organism going, and about how much they cost. We both arrived, in our different ways, at the view that trade-offs hold the key.

There is one other idea about the evolution of ageing that we should consider before we look into the question of how ageing occurs, and this is the notion that ageing happens not because of any trade-offs, but purely because natural selection is powerless against things that go wrong late in life.

If a harmful mutation should arise in a population of animals that has its effect at an age when most individuals will be long dead because of accidents, who cares? The idea that ageing might evolve simply because natural selection cannot prevent it was the brainchild of Peter Medawar in his inaugural lecture as professor of zoology at University College London in 1952. What Medawar proposed was basically the idea of genetic dust-beneath-the-cupboard.

The idea of genetic dust-beneath-the-cupboard works like this. Over the generations, new mutations happen all the time. Mutations which have harmful effects that reduce fitness get quickly eliminated by natural selection. How harmful a mutation may be is determined partly by how big an impact it has on the affected individual and partly by the age at which the effects of the mutation are felt. Harm is measured principally in terms of its effect on reproduction because this is what really determines evolutionary fitness.

Even a modest amount of harm will matter a lot if it affects very young individuals, because all of their reproduction is yet to come.

But a mutation that affects older individuals matters less because, in the first place, older individuals will have had some oppportunity to reproduce already, and in the second place, many of them might die from accidental causes before the age at which the mutation acts. The result is that natural selection will be good at sweeping out early-acting harmful mutations (dust-in-the-middle-of-the-floor) but poor at clearing out late-acting harmful mutations (dust-beneath-the-cupboard). The distinction is a question of how 'visible' the dust is to the cleansing action of the broom of natural selection.

When this process has gone on for a long while, as in a house that has not had the furniture moved for thorough cleaning, there will be a great deal of dust-beneath-the-cupboard. So long as no one moves the furniture, who is to know? But if the cupboard does get moved, the dust is there for all to see. This works as an analogy for ageing: if you remove an animal from the wild environment, where the mutations that act in the later years of its life really do not matter, and you transfer it to a protected environment, the late-acting mutations suddenly *do* matter.

There has been some debate about just how important a factor genetic dust-beneath-the-cupboard might be in the ageing process, and the evidence at the moment is ambiguous. It is likely that it counts for something, and certainly any accumulation of genetic dust-beneath-the-cupboard can only reinforce the disposability of the soma. However, the evidence that the soma is indeed disposable is much stronger.

We can now answer the question of why ageing occurs rather convincingly. Firstly, ageing is likely to happen because genes treat organisms as disposable: they invest enough in maintenance to enable the organism to get through its natural expectation of life in the wild environment in good shape, but more than this is a waste. Secondly, there may be design constraints that favour the interests of the organism in its youth at the expense of its long-term durability. Finally, natural selection in the wild is not much concerned with late-acting mutations, which may accumulate unchecked within the genome.

From our point of view, the problem is the same, whichever mechanism applies. Our rapid social and cultural evolution has outstripped our biological evolution. We have reduced our mortality from extrinsic hazards, only to expose all the more clearly our intrinsic mortality. In our battle against ageing's effects, we are up against the evolved design of our bodies and their systems of maintenance and repair. If we want to harness this insight to improve the quality of the later years of life, we need to know a great deal more about how the ageing process is actually played out. This is where we turn next.

CHAPTER 7

Cells in crisis

In my beginning is my end ...
... In my end is my beginning.

T.S. Eliot, 'East Coker'

The adult human body contains on average around a hundred thousand billion cells – that is, a one followed by fourteen zeros. Each cell, with the exception of red blood cells, is a complete unit with its own copy of the DNA that defines the individual and all the various chemical components that support life. Red blood cells do not have the DNA and can survive only a few months in the body, after which they are destroyed and replaced.

Cells come in various types. These range from the greatly elongated neurones of the brain and nervous system, which provide the wiring along which electrical nerve impulses are conducted, through the various specialised cells of liver, kidney, muscle and other organs, to the generalised cells that make up the so-called connective tissue. Connective tissue is what fills in the spaces between the more specialised cells and organs.

Some cell types are of a kind that is called post-mitotic – that is, they will not divide again. Brain and muscle cells belong to this category. Other cell types divide on a regular basis. Cell types that undergo regular divisions can usually be propagated, or cultured, outside the body.

If you wanted to grow cells, here is how you would do it. Take one small snip of skin – just a few cubic millimetres will be enough. Place in a sterilised dish or bottle made of plastic or glass. Cover with a pinkish liquid called growth medium. Leave in a warm cabinet set to the normal body temperature of 37°C with a carbon dioxide-enriched atmosphere to mimic conditions inside the body. After a few days you will have your culture of cells.

If you look down a microscope, you will see these cells forming a kind of halo around the original piece of skin. The cells are mostly fibroblasts, general-purpose cells that are found in most tissues of the body. A few of the cells will be more specialised skin cells like keratinocytes, but generally these special cells do not grow as well as fibroblasts – they get left behind.

If you turn up the power on the microscope, you will see that each fibroblast resembles a flattened octopus, its tentacles stretched out at either end and holding tight to the dish. Webs of protoplasm stretch between the tentacles. And the body of the octopus – the cell nucleus – is pressed down in the middle. Animal cells are in general not at all like the neat spheres and cylindrical rods of bacteria. Whereas the bacterial cell wall is rather rigid, the animal cell wall is elastic like an amoeba.

If you now wait a few days longer, you will find that the cells have multiplied by cell growth and division to cover the surface of the dish. Once this has happened, the cells will stop dividing and just sit there.

Just what makes them stop is not known, but it probably has to do with the same mechanism that tells cells in a healing wound to stop dividing once the wound is closed. The phenomenon is known as contact inhibition and it comes into play once a cell is surrounded on all sides by other cells. It is linked to the complex signals that each cell in the body uses to communicate with its neighbours. If our cells did not emit and receive these signals, we would carry on growing as spherical blobs, just as we did back in the earliest days of our life as an embryo.

If you want the cells to divide further, the trick is to space them out. The commonest way to do this is first to detach them by adding an enzyme called trypsin to the growth medium. Trypsin nibbles away at the glue that the cells use to hold tight to the dish. With the right amount of trypsin you can get the cells to detach completely and yet not be damaged. You then have a short window of time when the cells are floating around freely in the growth medium before they settle and attach themselves again.

What you must do now is suck up a fraction – typically a quarter or a half – of the growth medium containing the free-floating cells into a pipette, transfer it to a fresh dish, and top up with new growth medium. You can then leave the cells in the growth cabinet, or incubator, and in a little while they will settle down, grow and divide further until, after two or three days, contact inhibition stops them once more. You can then do this procedure all over again, and again, and again.

By this stage you have acquired the basic know-how of mammalian cell culture and, incidentally, gained a marketable skill. Cell cultures are widely used in medical research and in the pharmaceutical industry and they provide an alternative, in some contexts, to animal experimentation.

After several weeks of painstaking cell culture, studiously avoiding contamination with the myriad moulds and bacterial spores that float around in the air, you will look down the microscope and notice that something is going wrong. Your cells are sick, they are growing more slowly, they are failing to grip the dish as tightly as before, they are less regular in size.

In short, they are ageing. And soon they will stop growing altogether and eventually die. If you did not know that cells age and die in culture ('under glass' or 'in vitro' are common alternative terms), you would be worried that you were doing something wrong. You would be especially worried if everyone 'knew' this did not happen, and if everyone included your employer.

This brings us to an extraordinary and cautionary tale of how scientific progress can be subverted by the force of human personality and by the conservatism of dogma. In the final months of 1910, a charismatic and self-aggrandising French surgeon named Alexis Carrel began what would become a highly influential series of scientific reports on the culturing of animal cells outside the body.

A flurry of communications to the Société de Biologie in Paris was followed by a move to the Rockefeller Institute in New York, where Carrel worked between 1912 and 1939. Carrel's career was boosted when in 1912 he was awarded the Nobel Prize in physiology and medicine for his surgical studies, but as a supporter of fascism Carrel returned at the beginning of the Second World War to Nazi-occupied France, where he died in disgrace on 5 November 1944.

The most celebrated of Carrel's cell cultures was grown from the heart of a chicken embryo. The culture was started on 17 January 1912. It soon gained a press coverage that would be the envy of many a Hollywood starlet.

Carrel was a master in managing his media relations. When he discovered that the addition of a chick embryo extract to the growth medium increased the rate of growth of the cells in his chicken heart culture, word quickly went around that, when applied in Carrel's wound-healing studies, this treatment would dramatically accelerate the rate of tissue repair.[6] The *Philadelphia Star* of 16 January 1913 reported: 'While the good doctor doesn't come right out and say so, he leads us to believe that in future we will be quite exempt from all bodily injuries.'

In a little while it was not the growth rate but the longevity of Carrel's chicken cell culture that made the headlines. By the early 1920s the culture had been alive nearly ten years, long enough for New York's *The World* to remark on 12 June 1921 that, if all the cells had been retained, they could have formed a 'rooster ... big enough to cross the Atlantic in a stride; it would also be so monstrous that when perched on this mundane sphere, the World, it would look like a weathercock'.

Eventually the excitement subsided, but birthday announcements of the chicken cell culture continued to appear year after year. When Carrel returned to France in 1939, the culture remained in the care of his long-serving chief research assistant, Albert Ebeling, who took it with him to the Lederle laboratories of the

American Cyanamid Company, where it was finally discarded in 1946. By this time it had enjoyed 34 years of continuous growth.

At the same time that the public was revelling in stories of Carrel's amazing cells, in the quieter corridors of science it was accepted, on the basis of Carrel's published reports, that cells grown outside the body could live indefinitely and that the cause of ageing was to be found at some higher level of bodily organisation. The conventional wisdom ran that cell cultures were immortal *provided they were properly maintained.*

This acceptance came at a price: it contradicted experience. The situation was not unlike Hans Christian Andersen's classic tale of 'The Emperor's New Clothes'. With what we know now, there can be no doubt that culture after culture must have died out, but no scientist was ready to risk exposure as a failure.

Early rumblings of discontent can be detected in some papers of the 1950s, but it was not until 1961 that American cell biologists Leonard Hayflick and Paul Moorhead had the courage, determination and shrewdness to prove conventional wisdom wrong. In a landmark paper in the journal *Experimental Cell Research*, Hayflick and Moorhead demonstrated that normal human fibroblasts had finite replicative life spans. The cells would divide many times, but inevitably they would die out. Hayflick and Moorhead also made the audacious suggestion that the limited division of the cells might have something to do with the process of ageing.

The clever device that Hayflick and Moorhead used to silence their critics, who could otherwise have accused them of experimental sloppiness, was to grow a *mixed* culture containing 'old' female cells that had been through forty divisions in culture already, together with 'young' male cells that had been through only ten divisions. Male and female cells could be told apart by their different chromosomes.

The rationale of this experiment was that, if Hayflick and Moorhead's technique was at fault, perhaps because they allowed their cells to become infected, then the male and female cells would stop growing and die at the same time. But if the life span was intrinsic to the cells, then the 'old' female cells would stop dividing first, while the 'young' male cells would have thirty or so divisions still to go. It was the second outcome that was observed.

Even so, there were those who believed (a minority still do) that the limited growth of cells reflected some artifact, perhaps an inadequacy of the growth medium, causing the cells to run down gradually over many cell generations in culture. Hayflick and Moorhead's paper had a hard time before it was accepted. Indeed, the editor of the *Journal of Experimental Medicine*, the prestigious journal to which they first sent their manuscript, rejected it on the grounds that 'The largest fact to have come out from tissue culture in the last fifty years is that cells inherently capable of multiplying will do so indefinitely if supplied with the right [conditions].'

Once it was published, however, the result came to be confirmed in many other laboratories around the world and the 'Hayflick Limit', as the phenomenon of limited cell growth is now commonly called, has become an accepted feature in the culturing of normal cells.

What explains Carrel's result? We shall probably never know for sure, but three possible explanations have been suggested. The first possibility is that the chicken cell culture spontaneously 'immortalised'. The foregoing parts of this chapter have all dealt with *normal* cells. Cells grown from malignant, cancerous tissue are different. Such cells routinely give rise to so-called permanent cell lines. Permanent cell lines need not always start from tumours. Rodent cell cultures, grown from normal tissues of mice and rats, frequently undergo spontaneous immortalisation as they are cultured. Such cultures grow normally for a while, slow down as the normal cells reach their Hayflick Limit and stop dividing, and then speed up their growth again as the immortalised cells carry on growing and take over. What is more, such immortalised cells can often cause cancer if introduced back into the whole animal. The snag is that it is most unusual for a chicken cell culture to immortalise spontaneously, so although this explanation of Carrel's long-lived culture cannot be ruled out, it seems unlikely.

The second possibility is that the chick embryo extract used by Carrel and his team to stimulate the cells to grow more vigorously was inadvertently contaminated with fresh embryonic cells. In the early stages of the work, such an explanation might have been plausible, but later on this possible artifact was recognised and steps were taken to prevent it.

The likeliest explanation, and the simplest, is that Carrel's cells did in fact repeatedly die out and were deliberately restarted each time by his team of technicians, who were inculcated with the idea that such disasters were due to their own failings. By all accounts, Carrel was not the most forgiving of bosses.

There is some anecdotal evidence in support of this theory. Ralph Buchsbaum, a scientific visitor to the laboratory, arriving at a time when Carrel and Ebeling were both away, commented on the apparent poor health of the culture he was shown. Asked how such a culture could last so long, one of the laboratory technicians remarked: 'Well, Dr Carrel would be so upset if we lost the strain, we just add a few embryo cells now and then.'

The irony in this sad tale is that it had actually been predicted in 1881 by August Weismann that somatic cells should have finite replicative life spans, just as Hayflick and Moorhead discovered some 80 years later. Weismann died in 1914, just when Carrel's cultures were getting under way. Weismann's ideas about cellular ageing fell into neglect partly because Carrel's work showed him apparently wrong and partly, perhaps, because Weismann was an ardent nationalist who at the outbreak of the First World War lived long enough to renounce scientific honours bestowed upon him by Germany's enemies. In an era of rampant jingoism, this cannot have endeared his ideas to those in France, Great Britain or America.

Who knows what might have happened if fate had played these cards differently? Without Carrel's unfortunate intervention, it seems quite possible that the science of cellular ageing would have begun long before the 1960s.

Hayflick soon followed the discovery of the Limit by further

experiments designed to test whether the phenomenon had relevance to ageing. Others followed suit. It was quickly established
that the finite growth of the cultures was measured in terms of cell
divisions and not elapsed time. In other words, cultures restrained
from growth did not use up their life span; they did so only when
they divided. For this reason, the Hayflick Limit is often referred to
as replicative senescence, and the age of a culture is measured in
cumulative population doublings, or CPDs.

Advances in cryopreservation techniques soon allowed cells to be
frozen and stored at $-196°C$ in liquid nitrogen for long periods
of time; upon thawing the surviving cells were found not to have
aged at all. This greatly aided the husbandry of cell stocks and
permitted experiments to be repeated again and again with cells
from the same original strains.

Within a few years an important result was established, most
definitively by a team at the University of Washington in Seattle,
led by pathologist George Martin. It was found that the numbers of
cell divisions that human cultures could go through – their
replicative life spans – declined with the age of the cell donor. Here
was strong evidence to support Hayflick's contention that cell
ageing and human ageing were linked.

The longest replicative life spans were shown by foetal cells,
which could typically manage around 50–60 cumulative population
doublings. Cells from people in their nineties had lifespans
averaging around 25–30 CPDs. From birth, it was estimated that
each additional year of life knocked about one-fifth of a population
doubling from the tally, which suggests, incidentally, that in life
we do not renew our fibroblasts at a very great rate.[7]

It is worth reflecting on these numbers. An embryonic fibroblast
culture manages 50–60 cell population doublings. That gives
between a million billion and a billion billion cells from _one_
original cell. However, we start not just with one cell in our
embryo culture, but typically with around a million. This means
that, if all the cells were kept, there would eventually be between a
thousand billion billion and a million billion billion cells in the

final population. The average human adult contains roughly a hundred thousand billion cells. So the final mass of cells from the original embryo culture would be between ten million and one hundred million times bigger than you or me!

The fact that we have such awesome spare capacity is a conclusion that deserves greater recognition than it gets. Most of the research that is done on replicative senescence focuses on the end game, when the cells finally stop growing. But the arithmetic tells us that, if replicative senescence has anything to do with ageing, and I believe it has, then it cannot be the terminal collapse of the whole cell population that matters.

Some of the cells within a culture seem to reach the Limit sooner than others, either because they happen by chance to get through their 50–60 divisions more quickly than their companions, or because the Limit is not the same for each individual cell in the population. Either way, it means that some cells age while others are still young.

In 1995, Judith Campisi and her colleagues at the Lawrence Berkeley Laboratory near San Francisco discovered a protein that distinguishes old human cells, which have bumped up against their Hayflick Limits, from their younger counterparts. The protein, known as sen-beta-galactosidase, which shows up in the old cells as a blue stain when appropriate chemicals are added, has been used to demonstrate that, as we age, small but increasing numbers of senescent cells can be detected in our skin. This approach will be extremely valuable for tracking the contribution that the Limit really makes to the ageing process.

A natural question to ask is: why do the cells stop dividing; in other words, what causes the Limit? For many years, most of those working on the problem of the Hayflick Limit believed that cell ageing was programmed. If pressed further, most would link the idea of programmed cell ageing to the idea of programmed ageing of the organism as a whole. But programmed ageing of the organism does not make sense, as we saw in Chapter 5. So why do the cells stop dividing?

We find the field still polarised between those who think the cause is stochastic, a random accumulation of damage, and those who continue to favour the programme idea. All right, the latter group will now say, the Hayflick Limit is not programmed to cause ageing of the organism, but maybe the programme evolved to protect us against unlimited cell proliferation, otherwise known as cancer. If this is the case, then perhaps ageing, as mediated through the Hayflick Limit, is just the downside of a trade-off. And we saw in the previous chapter that trade-offs are important.

It is superficially rather an attractive idea that the Hayflick Limit does something useful. We like to find purpose in biology, and a mechanism that stops a tumour in its tracks is appealing. But just because we like purposive explanations, that does not mean they are right.

One of the first problems with the idea that the Hayflick Limit is an anti-cancer device is that in human cells it takes a long time to act. We have seen just how much cell expansion can occur before the cells hit the limit. Even adult cells, which have used up some of their divisions already, can still divide many times before the limit cuts in. During this time a tumour might be evolving all kinds of additional and nasty tricks, such as the ability to break free and spread via the bloodstream or lymphatic system to another part of the body. It might even acquire the mutations that will enable it to break free from the Limit. When it comes to killing cancer cells, the old adage that 'a stitch in time saves nine' has a powerful ring of truth.

The plausibility of the idea that the Hayflick Limit is there primarily to prevent cancers takes a further knock if we consider the differences between different animal species. If we grow fibroblasts from long- and short-lived animals, we find that the Hayflick Limit in the longer-lived animals is, itself, longer. This fits well with the connection between the Hayflick Limit and ageing, but what does it do for the possible connection with cancer?

A mouse has a Hayflick Limit of around 10–15 cell population doublings compared with the human limit of 50–60 CPDs. So you might at first think that the mouse would be better protected against cancer. Its cells are not able to divide as much before they stop.

Actually, mouse cells are much more cancer-prone than human cells. The lifetime risk of getting a cancer in a mouse is about the same as for a human, in spite of the fact that the mouse not only lives one-thirtieth as long, but is very much smaller and so has a smaller number of cells at risk. Measured as a rate per cell per day, the risk of a mouse cell becoming malignant is many thousands of times greater than the risk for a human cell.

For these reasons, the idea that the Hayflick Limit evolved primarily as an anti-cancer device and just happens to cause ageing as the downside of a trade-off is not, I believe, very convincing.

With the growing focus on molecular genetics, there has been a good deal of interest lately in identifying genes that get activated in old cells as they near the end of their replicative life spans. The idea is that these genes might tell us something about how the Hayflick Limit is controlled. One such gene, discovered by cell biologists Jim Smith and Olivia Pereira-Smith at Baylor College of Medicine in Houston, Texas, goes by the name of sdi-1, for 'senescence derived inhibitor'.[8]

Supporters of the idea that the Hayflick Limit is programmed, rather than the result of accumulated damage, welcomed the discovery of sdi-1 with open arms because the existence of such genes can be interpreted as evidence of the programme at work. However, this view may be mistaken.

We know that cells have a well-organised set of molecular controls that govern their progress through the cell cycle. In particular, there is a series of 'checkpoints' that each must successfully negotiate before it can proceed to division. Indeed, it would be surprising if things were otherwise, for division is quite the most taxing and dangerous task that the cell must perform.

The situation is like a passenger jet taxiing to the runway for take-off. The crew must run through a long list of pre-flight checks before they commit themselves and their passengers to hurtling down the runway at breakneck speed in the hope of soaring gracefully aloft before they reach the end. Suppose that a critical component of the aircraft has been damaged by an accident or malfunction. A red light will glow in the cockpit and the flight will be aborted. You and I know that it was the malfunction that caused the flight not to happen, but an observer who knew nothing about aeroplanes might be forgiven for thinking it was the red light itself that was the cause.

Often it is through recognising that truth lies beneath the surface that the door is opened to a better understanding. This, I assert, is why we so much need an evolutionary understanding of ageing to help us interpret a phenomenon like the Hayflick Limit.

There is plenty of evidence that random damage accumulates in cells as we get older. It is entirely plausible that this damage activates genetic mechanisms that repress cell divisions. This does not mean that the genetic mechanisms are there to *cause* ageing. Indeed, as Chapters 5 and 6 have shown, this is most unlikely. The genetic mechanisms probably evolved to protect young organisms from unsafe cell divisions. The fact that we see these mechanisms come into play more extensively in old age simply signifies that in old age they are needed more often, on account of more faults having accumulated.

Further evidence that randomness, if not necessarily damage, plays a role in cell ageing is evidenced if we look closely at cell cultures: they are extraordinarily heterogeneous. You might imagine that, as a culture grows towards the Hayflick Limit, the cells would march in unison. Nothing could be further from the truth.

A culture that has gone through, say, thirty population doublings might be halfway to the Limit and have thirty more populations to go. But as Jim Smith and others have shown, this does not mean that each and every one of the cells will divide thirty more times. If you were to isolate individual cells and measure how far each of them can grow, you would find that some had only a handful of divisions left, while others had more than thirty. It is even possible, as Robin Holliday and I showed back in 1975, to start with some cells in the culture that could grow indefinitely, but had a certain probability of giving rise to daughter cells that lost this power, and still end up with a culture that hit the Limit. The reason for this surprising conclusion, as we showed with a mathematical model, is that in the hothouse conditions of cell culture, the cells with indefinite life spans might be crowded out and eventually lost.

When we look at cell ageing in all its complexity, we find that we are up against a set of deep questions that we are still far from fully answering, but at least there are some promising leads. One of these deep questions is the extent to which ageing affects a rather special set of cells in the body, called 'stem' cells.

All of the cells in the body originate from the fertilised egg. The fertilised egg and the cells formed by its first few divisions are totipotent – that is, they have the potential to generate all the different cell types of the body. Stem cells are formed in the next stages of development.

While no longer totipotent, stem cells are pluripotent – that is, they have the potential to form multiple cell types. The best-known stem cells are those of the system for forming blood cells, the so-called haemopoietic stem cells, which are found in bone marrow. The purpose of a bone marrow transplant is to transfer haemopoietic stem cells. You might want to do this as a treatment for leukemia, a cancer of the blood-forming cells. The treatment involves destroying the malignant cells by irradiation and then reconstituting the system with fresh stem cells from a tissue-matched donor. The treatment works because stem cells have great regenerative powers; a small number of stem cells can reconstitute the entire blood cell population.

Although stem cells have the ability to generate large numbers of cells, they do not themselves divide at a great rate. In fact, they divide infrequently, but they give rise to enormous numbers of clonal descendants. (In the terminology of cell biology, the cells produced by division from a common cellular ancestor are called a clone. Also, by convention the two cells produced from a single cell division are known as daughter cells, even if the cells concerned happen to be male.) Remember that cell division is a doubling game, so one daughter cell from a stem cell will go on to form a thousand progeny by the time it has divided just ten times. Stem cells sometimes divide to form two new stem cells, but usually they divide asymmetrically, one daughter cell remaining a stem cell, the other starting down the pathway to becoming a specialist – for example, a white blood cell – and losing the pluripotency of its parent. In this way, the organism can keep topping up the different kinds of cell that are needed.

It is a neat system, but do stem cells themselves age and die? In the early 1970s an attempt was made to measure just how long a haemopoietic stem cell pool could be made to last by carrying out serial bone marrow transplants in mice. In these experiments, bone marrow was transplanted from a donor mouse to a recipient, which then became a donor in turn. The transplants worked for several successive transfers, but eventually failed to take, or gave rise to leukaemias. Could this be because the stem cells aged?

As far as the haemopoietic stem cells are concerned, we do not yet know the answer. One of the most important functions of the haemopoietic stem cells is to keep supplying the B and T lymphocytes, which are the cells of the immune system that fight infection. In the course of normal ageing, there is no suggestion that we run out of these cells, but it has been suggested that the decline in the numbers of one category of T lymphocytes (the CD4 positive cells) in people with AIDS might be due to exhaustion of the regenerative capacity of the immune system. There is massive destruction of this particular category of T lymphocytes caused by their infection with HIV, the human immunodeficiency virus, and the renewal of these cells places an exceptional demand on the haemopoietic system. Whether this is a plausible explanation of AIDS remains unclear, but signs of ageing in stem cells are beginning to be discovered elsewhere in the body.

In the lining of the gut, there is an elegant stem cell system, which is found in little pocket-like structures that go by the fantastical name of the crypts of Lieberkühn. To find the crypts of Lieberkühn we need to examine the gut wall rather closely.

If you have ever looked closely at the lining of an animal's stomach, perhaps among the offal at a butcher's, you will perhaps have noticed that the inner side is furry. The furriness comes from the fact that the inner wall of the stomach is covered with fine hair-like projections called villi (singular: villus). So is the wall of the gut. Unlike the dead keratin of real hair, the villi are made up of living cells. The crypts are found between the bases of the villi. Near the bottom of each crypt in the small intestine of a mouse sits a group of between four and sixteen stem cells. These divide to give rise to 'transit' cells, which migrate, dividing as they go, up the walls of the crypt, across the gut surface, and then up the sides of the villi, where eventually they get shed into the gut at the rate of about 1,400 cells per villus per day (see Figure 7.1).

The beauty of the gut stem cell system is that the stem cells themselves can be identified by their position within the tissue. This makes them much easier to study than most other kinds of stem cell. One of the things that is known about gut stem cells is that they maintain themselves in a rather special way. When there are too many of them or when one is damaged – for example, by irradiation – a stem cell is likely to commit suicide! This, incidentally, is why the gut is one of the first organs in the body to show the effects of radiation sickness.

Cell suicide, or apoptosis, in which a cell actively destroys itself, was discovered in the early 1970s and has come to be seen as an extremely important part of being a multicellular organism. In animals, apoptosis plays an essential part in the shaping of our bodies during development. For example, the gaps between our fingers and our toes are formed through its action. Apoptosis also occurs in plants, notably in the mechanisms that cause leaves of deciduous trees to separate and drop in the autumn, changing colour before they do so. This striking effect changes entire landscapes. For this reason, apoptosis has the distinction of being

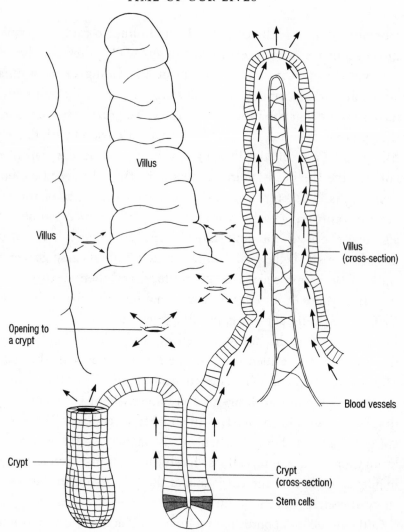

Villus

Villus

Villus
(cross-section)

Opening to
a crypt

Blood vessels

Crypt

Crypt
(cross-section)

Stem cells

FIGURE 7.1
The cellular organisation of the gut wall. The inner surface of the gut
has many tiny projections, called villi. Between the villi are openings
which lead to flask-shaped structures called crypts. Near to the bottom
of each crypt is a small number of stem cells. The stem cells give rise
to daughter cells which divide rapidly and move up the crypt as they
do so. After leaving the crypt, the cells migrate over the gut surface
and up the sides of the villi. They are shed when they reach the villus
tip. From Potten & Loeffler, *Development*, Vol. 110, pp 100–1020, with
permission.

perhaps the only cell-mediated process whose effects can be seen from space.[9]

Apoptosis also seems to be important in keeping us alive. Damage that leads to mutation in a stem cell is potentially a very dangerous thing because it results in lots of wrong cells being formed, and confers an increased risk of malignancy. So if a stem cell senses that it is has been harmed in this way, it activates a rather closely controlled death mechanism, it disappears, and its place is taken by the offspring of a neighbouring stem cell that remains intact.

In principle, the stem cell population of each crypt might keep itself going more or less indefinitely through the purifying action of apoptosis. However, recent work done in collaboration between my laboratory and that of Chris Potten of the Paterson Institute for Cancer Research, Manchester, who has played a leading role in elucidating the properties of the gut stem cell system, shows that this is not the case.

The gut wall changes with age and within the crypts the stem cells are more liable to undergo apoptosis when challenged with low doses of damage. This suggests that they may have accumulated some damage already, bringing them closer than young cells to the threshold needed to activate the death mechanism.

When apoptosis was first discovered, there were many in the field of ageing research who hailed it as the agent of death, which those who believed in the programme theory suggested existed. It now seems far more likely that apoptosis is primarily an agent not of death, but of survival. Just as we saw earlier with the gene *sdi-1*, the key question is one of ultimate cause and effect. There is no evidence that ageing is driven by apoptosis, even though apoptosis does occur more readily with age in some tissues, probably because damage accumulates. In the nematode worm *Caenorhabditis elegans*, mutations have been found that inactivate the apoptosis mechanism. The mutant worms have some developmental abnormalities, but their life spans are not extended.

So far we have looked at the changes that occur with age in dividing cells like fibroblasts, and in tissues where stem cells are important, but what of cells that no longer divide, such as the neurones of the brain?

These post-mitotic cells, as non-dividers are called, are undoubtedly involved in ageing and there is plenty of evidence, as we shall see in later chapters, that they accumulate damage as the body ages. When damage overwhelms such cells, they die and their numbers decline. Sometimes they die by apoptosis, which is a neat and tidy death and probably happens to avoid leaving a cell corpse that might trigger an inflammatory reaction. But sometimes they die the messy way, and local inflammation may occur.

One of the most poignant examples of cell death is seen in the brain following a stroke, when the blood supply to a part of the brain is blocked off by a clot. Deprived of oxygen, the cells that are immediately affected will die. However, much of the damage done in a stroke occurs through secondary, apoptotic deaths in the surrounding areas of the brain. Just why these cells die is still not clear – they are not obviously damaged. It may be that recovering from a stroke is just one of those things that natural selection has not prepared us for, and that the cells die because the suicide signal emitted from the damaged cells spreads unchecked. In this case, blocking the signals that trigger cell death may be beneficial and research is under way to attempt it.

Ageing brings cells to the point of crisis in many ways, and nothing underlines this fact better than the rare genetic disorder known as Werner's syndrome. Werner's syndrome affects around 1 in 100,000 individuals. It results in a pathological race through the sufferer's life span at twice the normal speed. Individuals affected by Werner's syndrome die usually in their thirties or forties, with multiple conditions such as heart disease, cancer, osteoporosis and immune dysfunction.

It seems that whatever is going wrong in Werner's syndrome is linked with faster ageing of their dividing cells. When George Martin and his colleagues in Seattle did their pioneering work which showed that the Hayflick Limit in human fibroblast cultures declined with the age of the cell donor, they included three samples from individuals with Werner's syndrome. They found that the Hayflick Limit of the cells from Werner's syndrome patients was very much less than that of cells from normal individuals without the genetic abnormality.

This discovery also explains a curious feature of Werner's syndrome, which is that the brain and muscles are largely unaffected. People with Werner's syndrome do not suffer early onset of Alzheimer's disease, for example. It seems that the genetic defect that has such a striking effect on cell division does not affect non-dividing cells to anything like the same extent.

In 1996, an explanation for this big difference between dividing and non-dividing cells came to light. A team of geneticists based in Seattle, including George Martin himself, identified the gene whose mutation is responsible for Werner's syndrome. The gene provides the instructions to make a protein of a kind known as a helicase.

Helicases are proteins that unwind the DNA double helix in readiness for copying and repair. Dividing cells must copy their DNA each time they divide, whereas non-dividing cells have no such requirement. The discovery that a broken helicase appears responsible for Werner's syndrome fitted the biological and clinical observations very well.

The fact that a defect in the machinery for copying and repairing DNA had the effect of accelerating so many features of the ageing process was exciting indeed. We saw in Chapter 6 how the disposable soma theory supports the idea that normal ageing is caused by an accumulation of cellular damage, and for many years there has been interest in the idea that damage to DNA is a key part of the story. The Werner's syndrome breakthrough was confirmation of the tantalising link between the ageing of cells and the accumulation of faulty molecules.

CHAPTER 8

Molecules and mistakes

Turning and turning in the widening gyre
The falcon cannot hear the falconer;
Things fall apart; the centre cannot hold;
Mere anarchy is loosed upon the world.

W.B. Yeats, 'The Second Coming'

At a little after 8 o'clock on the morning of 6 August 1945, a casual observer might have looked up at a distant aeroplane and, if the observer's eyes were sharp enough, seen a small object detach itself from its fuselage and fall earthwards. A short while later, the small object became the heart of a searing flash of light, unleashing a wave of energy so intense that the observer was scorched beyond recognition. The aeroplane was the Enola Gay, the place Hiroshima.

The atomic bombings at Hiroshima and Nagasaki were the culmination of feverish research by physicists, driven by the need to end the nightmare of the Second World War and sustained by the intellectual excitement of new discoveries about the fundamental forces binding together the atoms of our universe. The atomic bombings also signalled for nuclear physicists the end of an era of innocence.

In the years following 1945, a number of gifted scientists began to turn their attention from the uneasy world of nuclear physics towards biology, where the application of physical tools, such as the use of X-rays to probe the structure of biological molecules, was beginning to throw light on some of the basic puzzles of life. From this union of biology and physics, together with biochemistry, has grown the enormously successful discipline of molecular biology.

One of these physicists was Leo Szilard, a Hungarian who bore more than a little responsibility for the development of the atomic bomb. It was Szilard who in London in 1933, having left his work in Germany two months after Hitler was appointed chancellor, first had the idea of the atomic chain reaction. And it was Szilard who, having moved on to the United States, teamed up with fellow Hungarian physicist refugees Eugene Wigner and Edward Teller, to persuade Albert Einstein to write a letter to United States President F.D. Roosevelt alerting him to the potential of the atom bomb and the danger that Nazi Germany might build it first. Indeed, Szilard drafted the letter.[10]

In 1959, Szilard brought his knowledge of radiation to bear on the question of ageing. Szilard's reasoning was this: radiation is known to be an effective way to shorten the life span of animals; it is also known to induce mutations; therefore, could mutations in the somatic cells of the body be the cause of ageing? Szilard's idea was among the first of a number of theories that have suggested that ageing is the result of damage to the molecular structures on which life depends. In this chapter we will look into these ideas, but first we need to remind ourselves how living systems operate when all goes well.

Inside each cell of your body an amazing variety of chemical reactions is taking place. If you think of a chemical factory or oil refinery, with its bewildering complexity of pipes, valves, tanks and cables, you will have some idea of the sort of thing that is happening, except that inside the cell many of the ingredients simply slosh around together. A special compartment houses the nucleus, which contains most of your genes. The nucleus is separated from the rest of the cell by a membrane, which is something like a plastic bag with holes in it. The material of the bag is rather special, however, and the holes are cunningly designed to regulate what goes in and what comes out.

Outside the nucleus but within the main compartment of the cell, known as the cytoplasm, are various other bags full of things. Some of these are organelles, which are more or less permanent companions to the nucleus. The mitochondria (singular: mito-chondrion) are a cluster of organelles that constitute the power sources of the cell and provide energy to fuel the other reactions that need to take place elsewhere. Each mitochondrion has its own special set of genes, which are additional to the genes of the nucleus.[11]

As well as the mitochondria and other minor organelles, there is a miscellaneous collection of temporary bags that carry things inwards from the outer membrane that surrounds the whole cell, and which take things out. Amidst the intracellular bustle, there extend various strands and fibres that hold bits together or pull them apart. Finally, on the outside of the cell membrane is a collection of structures that link the cell to other cells and which carry the cell's chemical badges of identity, so that the other cells know how to respond appropriately when their paths happen to cross (see Figure 8.1).

All this is a pretty impressive degree of sophistication and yet so far we are only talking about appearances, the sort of detail you can actually see if you have an electron microscope. We have not even touched on the basic chemistry, and we will skip much of the detail for it is truly complex. Various manufacturers of laboratory chemicals obligingly produce wallcharts of the known chemical pathways that occur within cells. These charts make a map of bus routes even in a major city like New York or London look simple by comparison. A biochemical pathway is a diagram that contains information of the following sort: A makes B into C, which combined with D makes E into F and G, and so on, where A–G are elaborate chemical compounds with names like pyruvate carboxy-lase and 1,3-diphosphoglycerate.

One might well worry that this complexity will go on and on in an infinite regress as we look more closely into the cell. Fortu-nately, however, there is a set of processes at the heart of all living

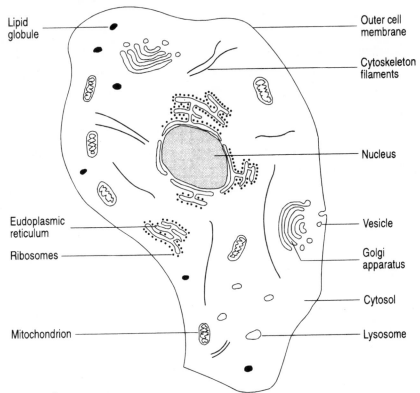

Lipid globule

Outer cell membrane

Cytoskeleton filaments

Nucleus

Eudoplasmic reticulum

Ribosomes

Vesicle

Golgi apparatus

Cytosol

Mitochondrion

Lysosome

FIGURE 8.1

A typical human cell. The nucleus, which contains the chromosomes, is separated from the rest of the cell by a double membrane. Immediately outside the nucleus is the endoplasmic reticulum which consists of flattened sheets and sacs, which may also be found detached from the nucleus. On the outer surface of parts of the endoplasmic reticulum are found the ribosomes which play a key role in the manufacture of proteins. Mitochondria are the power units of the cell, harnessing energy by combining oxygen with food molecules. Lysosomes are membrane bags containing powerful enzymes to digest and break down unwanted materials in the cell. Lipid globules contain essential fats. The Golgi apparatus is a system of stacked, flattened bags of membranes involving in packaging, sorting and delivering molecules from one part of the cell to another. Smaller membrane bags, called vesicles, carry materials to and from the Golgi apparatus and to and from the outer cell membrane. Throughout the cell a network of various strands and filaments makes up a structural framework called the cytoskeleton. The general fluid found inside the cell, but outside the nucleus, is called the cytosol. Some specialised cells also have long extensions to connect with other, distant cells in the body, or special junctions between adjacent cells; these features are not shown here.

cells that is stunningly elegant in its simplicity, yet amazingly rich in the diversity of life forms that can be made from it. This set of processes is the stuff of molecular biology, and its elucidation during the 1950s and 1960s has been one of the outstanding achievements of science.

One molecule takes pride of place. This molecule is deoxyribonucleic acid, better known as DNA. DNA does not have a fixed size and form like a molecule of water. Rather, it is constructed according to a particular set of rules and it takes a countless variety of individual forms, which all have this in common: all are composed of a chain of four different kinds of link. Each has a part in common with the other links, which makes for a uniform structure, and a part that is different to the others. The parts that are different are the nucleic acids: adenine (A), cytosine (C), guanine (G) and thymine (T). Because the structure is regular, all that we need in order to describe it is the *sequence* of different links. Thus, a typical DNA sequence might begin AACTGCG-TACCGTTG ... and continue in this vein for many hundreds or thousands of further letters.

A very special property of DNA is its capacity to permit an exact copy of itself to be made. Most of the time DNA is not in fact a single chain. It is two complementary chains that are twined around each other like a piece of two-ply wool. This is the celebrated double-helix structure, discovered in 1953 by James Watson and Francis Crick, working at the Cavendish Laboratory in Cambridge, England. In a sentence of epic understatement, the paper in *Nature* announcing their discovery concluded: 'It has not escaped our notice that the specific pairing we have postulated immediately suggests a possible copying mechanism for the genetic material.'

The crucial feature of the double-helical structure of DNA lies in the complementarity of the pairs of nucleic acids in the two opposite strands. The structure fits if an A in one strand is positioned opposite a T in the other strand, or if a C is opposite a G, but not otherwise. What this means is that, if the helix is unwound, then each of the single-stranded chains can be used as a model, or template, from which two new double helices can be formed that are exact replicas of the original. We can see this with a simple example.

Suppose we have a short double helix in which one strand has the sequence ATGACC. The complementary strand has the sequence TACTGG, because T must match A, and G must match C. We can write the two strands one above the other thus:

Strand 1: ATGACC

Double-stranded DNA | | | | | |
                          | | | | | |

Strand 2: TACTGG

where the lines joining the corresponding nucleic acids signify chemical bonds that hold the strands together. If we now gently break these bonds and separate the two strands, we have two single-stranded DNA molecules where the nucleic acids in each strand are looking for chemical partners:

Single-stranded DNA   Strand 1: ATGACC
                                    | | | | | |

                                    | | | | | |
Single-stranded DNA   Strand 2: TACTGG

New nucleic acids are supplied by the cell to complement the single-stranded molecules, and when the process is complete we have two double-stranded DNA molecules, which are identical to the double-stranded molecule we started with:

Strand 1: ATGACC

Double-stranded DNA | | | | | |
                                       | | | | | |

Strand 2: TACTGG

Strand 1: ATGACC

Double-stranded DNA | | | | | |
                                       | | | | | |

Strand 2: TACTGG

Bold type distinguishes the newly added nucleic acids from the nucleic acids that were part of the original molecule.

The beauty of this is that it explains how genes are passed faithfully from parent to child, for DNA is the molecule that genes are made of. But DNA is replicated not only for the purpose of passing genes on to the next generation. DNA is also replicated each time a cell divides, since our genes must be put to work in all cells of the body.

The way most genes work is by providing information to make proteins. This information is encoded in the DNA sequence like a message in a strand of tickertape. To make a protein the cell translates the gene sequence, written in the four-letter DNA alphabet, into a new sequence in the twenty-letter alphabet of amino acids. Just as DNA is made up of a chain of nucleic acids, proteins are made up of chains of amino acids. A protein is typically between 50 and 500 amino acids long (although some are shorter and others are longer).

Proteins are both building blocks in the cell (and outside it too)[12] and also tools. Each protein molecule therefore needs to be of the right shape and size if it is to function properly, and the different proteins come in a great variety of different shapes and sizes. The number of different permutations that can be made from the alphabet of twenty amino acids provides for a huge diversity of form. For a modestly sized protein just fifty amino acids long, there are a staggering 20^{50} alternative sequences – that is,

112,589,990,684,262,400,000 billion billion billion billion billion different possibilities!

The making of proteins, which operates as follows, is a miracle of micro-engineering. A working copy of the DNA sequence is made by 'transcribing' a stretch of DNA into a stretch of the chemically similar ribonucleic acid, or RNA. In fact, a gene may be stored in the DNA sequence in a number of separate parts, which get spliced together at this stage to give a continuous message, much as one might edit a piece of audio tape or cine film.

The working copy, which is called messenger RNA, or mRNA for short, gets transported to a molecular engine called a ribosome. Meanwhile, a set of enzymes called synthetases are busy sticking single amino acids on to the ends of curious little nucleic acid chains, which have been folded back on themselves to form clover leaf structures. These clover leaves are known as transfer RNAs, or tRNAs for short, and they have the ability to carry an amino acid at one end, while at the other end they expose a short stretch of three nucleic acids that comprise something called an anticodon (See Figure 8.2).

Given that we have an anticodon, it will be no surprise that there also exists a codon. To find the codon we go to the messenger RNA, which is now waiting at the ribosome. What the ribosome has done is position the mRNA so that the first three nucleic acids of the gene sequence are exposed at a special point on the ribosome, called the acceptor site. This triplet of nucleic acids is the codon. Poised nearby are a further set of enzymes that will serve as the delivery team to facilitate the birth of the new protein. Labour is about to begin.

Like human childbirth, the birth of a protein begins with waiting – waiting in this case while tRNAs clutching their amino acid cargoes drift up against the codon that is exposed at the acceptor site. If the anticodon of the tRNA fails to match, the luckless tRNA drifts away again, but if the match between codon and anticodon is good, the tRNA sticks and the ribosome springs into action.

The mRNA clicks along by three nucleic acids, dragging the

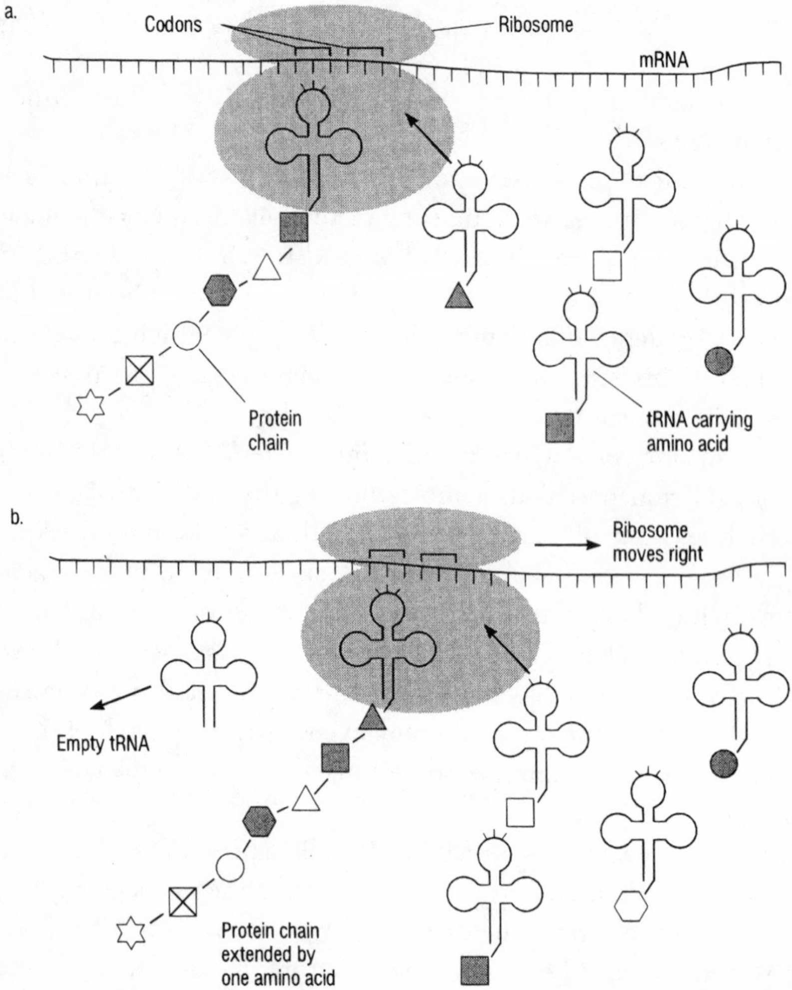

FIGURE 8.2

How proteins are made. The ribosome is attached to the messenger RNA
and holds two codons of the message, i.e. six letters of the RNA sequence.
To the right of the ribosome wait various transfer RNAs (tRNAs) each
carrying a single amino acid, represented by the different symbols at the
bottom of the clover-leaf tRNAs. At the left-hand codon on the ribosome
sits the transfer RNA to which the growing protein chain is currently
attached. In (a) the right hand codon on the ribosome is empty waiting for a
matching tRNA to approach it. The matching tRNA attaches to the empty
codon, the protein chain is transferred to join the amino acid on the newly
arrived tRNA. The ribosome moves rightwards along the messenger RNA by
a distance of one codon. The tRNA from which the protein was transferred
now carries no amino acid; it falls off the ribosome to go and pick up
another amino acid. In (b) the growing protein chain can be seen to have
grown by one amino acid, and the cycle is ready to repeat itself.

tRNA and its amino acid cargo to a new site on the ribosome, and now a new empty codon sits at the acceptor site. Another wait, a few more failed attempts at anticodon–codon matching, and it happens again. The second codon is matched by the anticodon of a second tRNA, and the mRNA moves a second time. But now a new step in the dance of protein synthesis occurs. The amino acid from the first tRNA gets linked to the amino acid that is still held by the second tRNA, and the first tRNA gets nudged away. The empty first tRNA drifts off to find a synthetase and get loaded up again with a new amino acid. And so it goes on.

Each fresh anticodon-codon match triggers the following actions: the mRNA moves along to expose a new codon, the last-but-one tRNA attaches the growing protein chain to the amino acid on the latest tRNA to arrive, and then the last-but-one tRNA, which is now empty, gets nudged away. The process ends when a special kind of codon, called a stop codon, is reached and the newborn protein is released to begin a life of its own.

Some precocious little proteins, we might call them urchin proteins, can handle the next stage of their lives all by themselves. They fold up into their correct three-dimensional structures and off they go. In Chapter 6, I likened a protein to a necklace made of twenty different kinds of bead. A charm bracelet might be a better analogy. Each kind of amino acid has a link element that is the same, and something called a side-chain that differs from the other kinds in respect of size, shape, electrical charge, affinity for water, and so on. But proteins do not drift around the cell as floppy chains. They need to fold into compact globular structures. Many proteins need help to fold correctly. This help comes from a special set of protein molecules called chaperones.

In time, proteins get damaged or they are not needed any more because the cell has switched to doing something else. Damaged and redundant proteins get tagged for disposal and carted off to a special kind of bag within the cell called a lysosome, where they are broken down by proteolytic enzymes that cut them into pieces, and are recycled.

Neat, eh? You haven't heard the half of it. The picture I have described of life inside the cell is the truth and nothing but the truth, but it is not the whole truth. The whole truth is a great deal more exciting, for all that I have described happens against a background of chaos and mayhem. No one is directing operations, they just happen, and mistakes happen too.

Take DNA, for example. The replication of DNA is astoundingly accurate: only about one in a billion letters gets copied wrong.* But this is because the copying enzymes are ferocious proofreaders as well. Lots of mistakes get made on the copying enzymes' first pass along the DNA strands, but they back up and fix them. And DNA gets damaged all the time. All kinds of things damage DNA. A veritable army of DNA repair enzymes works away, snipping out damage and stitching in correct sequence. In spite of all this activity, mutations accumulate in somatic cells as we live our lives. No molecular system is perfect. The DNA sequence gradually gets corrupted.

Not only does the sequence get corrupted by mistakes in copying and repair, but the control of gene expression – how genes are activated – can go awry. The cell needs ways to tell which genes should be on and which should be off. One of the ways this is done is by the use of little chemical tags in a process called DNA methylation. This is a bit like putting 'Post-It' notes in the pages of a large document to mark places you need to go back to.

Imagine now that you were to photocopy the document. You take the Post-It notes off the pages to copy them, and if you put them back at once, you keep the secondary information that they carry; you can even stick new Post-It notes on the pages of the new photocopy without needing to read the text again. But if you delay putting a Post-It note back and copy more pages before you realise your mistake, you are in a mess.

There is evidence that, as we age, our somatic cells lose DNA

* This is like copying out the entire *Encyclopaedia Britannica* and getting just a few letters wrong.

methylation patterns in much this way. As a result, they turn genes on that should be off, and vice versa. Such changes are called epimutations because they do not change the DNA sequence itself, but the effect is much the same as a mutation.

A rather special process affects the ends of chromosomes, where the long DNA strands terminate. The DNA copying machinery has a lot of difficulty with the ends of the DNA helices, which are protected by special structures called telomeres. Telomeres are sometimes likened to the plastic tips that keep the ends of shoelaces from unravelling. An enzyme called telomerase keeps telomeres in good shape when germ cells replicate, but for some reason telomerase is not active in most human somatic cells and telomeres get shorter and shorter as cells divide until, some suggest, they reach a critical length and stop the cell in its tracks. This is not a mistake in the sense of DNA copying errors, but it may have a special relevance to cancer, as we shall see in Chapter 10.

DNA can also be the agent of its own undoing because of 'selfish' replication. Selfish DNA elements, known as transposable elements, are akin to retroviruses like herpes and HIV in that they insert their own genetic sequence into the sequence of the host cell, causing mutations in the process. Unlike a virus, transposable elements cannot move from cell to cell, but they can copy themselves inside the cell, and they will do so if they can. There is clear evidence that a form of senescence in a species of fungus called *Podospora anserina* happens because of the selfish replication of a DNA transposable element. Human cells have transposable elements too, but it is not known whether their selfish replication contributes to ageing.

Proteins have their own perils. Protein synthesis is inherently much trickier than DNA replication because of all the extra steps that are involved, and because instead of just four nucleic acids to choose from, the protein synthesising machinery has to choose from a menu of twenty amino acids.

Mistakes in protein synthesis happen all the time. Sometimes the wrong anticodon sticks to the codon just long enough to trigger the ribosome, and the wrong amino acid joins the chain. Sometimes a synthetase attaches the wrong amino acid to a tRNA, so the anticodon matches correctly, but the wrong amino acid gets incorporated just the same. Sometimes, the ribosome fumbles its work and the protein drops off before it is finished. The upshot is that the error rate in protein synthesis can be as high as one wrong amino acid in every thousand. This is still an impressive level of accuracy, but an error rate of one in a thousand means that four out of every ten proteins that are 500 amino acids long get made with at least one mistake in them.

Errors in protein synthesis can lead to all sorts of trouble down the line. This is because proteins themselves are the actual movers and shakers of life inside the cell. Suppose you make a mistake in a protein that regulates gene action, and suppose that this mistake causes the protein to stick to the DNA molecule too tightly. If the faulty protein does not detach when it should, it can screw up the making of a messenger RNA, for example. In some cases it might even kill the cell. Other kinds of mistake generate more mistakes in their turn. A mistake in making an enzyme that copies DNA, or in a synthetase that sticks amino acids to tRNAs, might have this effect. These are the kinds of mistake that might contribute to the process of error catastrophe that I had been pondering when I first had the idea of the disposable soma theory (see Chapter 6).

We are warm-blooded, we eat hot food and we live on a planet where the sun shines. All of this is a good thing, but our proteins are not always so happy about it. Proteins can fold incorrectly, or they can flip out of the right fold on account of the molecular vibrations that occur when they are exposed to heat. In fact, a special family of enzymes – called heat-shock proteins – has evolved to clear up the mess when proteins get too hot and shake themselves into wrong configurations. It turns out that the heat-shock proteins also deal with damage that arises from a whole gamut of other stresses too, but it is significant that they were discovered because they allow cells to survive a sudden burst of heat. We all have heat-shock proteins. So do mice, worms, flies and even bacteria. They are very important.

Like DNA with its 'Post-It' methylation tags, proteins have chemical tags too. In fact, there is a lot of post-synthesis processing that happens to proteins. A protein can be subjected to phosphory-lation, ribosylation, acetylation, glycosylation and more besides. All of these '-ylations' involve chemical tagging. They affect how the protein works and where, inside or outside the cell, it will end up. Any of these modifications can be done incorrectly, or remain undone when it should be done. All of this provides for an impressive repertoire of mistakes.

Clearing up the mess when proteins are made wrongly, are damaged or just flip into incorrect folds is the job of the chaperones and scavengers. Chaperones may try to restore a deranged protein to its proper condition.[13] Failing this, the recalcitrant faulty protein receives the attention of a special squad of molecular heavies called protein scavengers, which, in the language of political assassina-tion, will deal with it with extreme prejudice. The trouble with the scavengers is that, like assassins, they may hit the wrong target. You do not get extensive scavenging without a certain amount of death by 'friendly fire'; plenty of perfectly good proteins get wasted that way.

The point of dwelling on all the unfortunate things that *can* go wrong with proteins is that they *do*, and they do so quite often. Bad proteins accumulate with 'normal' ageing, and they are implicated in a range of age-associated diseases as well, as we shall see in Chapter 9.

After reading this list of all the mishaps that can occur within your cells, you might be feeling that surely there is nothing more that can harm you. You would be mistaken. There is oxygen. The good citizens of smog-burdened cities like Los Angeles, Tokyo and Athens will not need me to tell them that a breath of air is not always good for you. Sadly, this is also true for the rest of us. And the reason is to be found in the two-edged properties of that essential chemical element, oxygen, without which we would rapidly die.

The trouble is that oxygen doubles as a deadly poison. Life as we know it could not have started at all if the primitive atmosphere of the earth had been as rich in oxygen as it is now. Oxygen does its damage by oxidising, which is what makes a fire burn, a car rust and butter go rancid. The energy we depend on for life is released when oxygen is used in a chemical reaction called oxidative phosphorylation. This is, if you like, a controlled form of burning.

Out of every 100 oxygen molecules that we take in, two or three slip away to become free radicals. Free radicals are not wholly bad for you – your immune system uses them to kill invaders – but they do an impressive amount of damage. What is more, they are indiscriminate in their damage; they zap DNA, proteins, membranes and almost anything else that gets in their way. Free radicals are intracellular vandals.

It is estimated that each cell of your body experiences as many as 10,000 hits by free radicals to its DNA every day! You should thank your lucky stars for your DNA repair systems or you would be in desperate trouble. And the situation would be a whole lot worse if your cells did not mount an impressive battery of antioxidant defences. Chief among these defences are antioxidant enzymes such as superoxide dismutase, catalase and glutathione peroxidase, which are aided by certain vitamins, notably C and E.

When a molecule of vitamin C encounters a free radical, it becomes oxidised and thereby renders the free radical innocuous. The oxidised vitamin C molecule then gets restored to its non-oxidised state by an enzyme called vitamin C reductase. It is like a boxer who goes into the ring, takes a hit to his jaw, goes to his corner to recover, and then does it all over again.

There is compelling evidence that oxidative damage accumulates in our cells and tissues as we live our lives, and there are many who believe that this accumulation of damage contributes to the ageing process.

If you are a molecule navigating around the inside of a cell and you want to avoid trouble from free radicals, you will be well advised to keep clear of the mitochondria. The mitochondria are where oxidative phosphorylation takes place and where the density of free radicals is greatest. Not surprisingly, mitochondria take a lot of hits.

Because mitochondria are continually exposed to high levels of radical damage, the membranes of the mitochondria are loaded with vitamin E. Membranes consist of fatty molecules in which a free radical can start a kind of oxidative chain reaction. Vitamin E quenches the spread of this chain reaction in much the same way that the control rods in a nuclear reactor quench a nuclear chain reaction.

You will recall that the mitochondria are organelles that once originated as free living cells in their own right. A part of the legacy of this independent life is that mitochondria have their own DNA, which codes for the enzymes needed for oxidative phosphorylation. But mitochondrial DNA is relatively unprotected because mitochondria have limited DNA repair capability. And mitochondrial DNA is in the front line of free radical attack. The upshot is that mitochondrial DNA suffers a lot of mutation. The situation is a little less catastrophic than it sounds because each cell has lots of mitochondria, and the mitochondria get replicated independently of the normal cell cycle. (Even in a post-mitotic cell, like a neurone, the mitochondria are continually being recycled and renewed.) Nevertheless, there is evidence that the level of mitochondrial mutations increases with age, especially in the brain and in muscle.

The ground we have covered in this chapter shows just how much can – and does – go wrong inside the cells of our bodies as we live our lives. In some ways, it makes it all the more remarkable that we live as long as we do. When life's vicissitudes seem too much to bear, it is worth remembering just how amazing each of us really is.

Something I have not mentioned explicitly so far is that the various maintenance and repair systems – the DNA repair enzymes, heat-shock proteins, antioxidant defences and so on – all cost energy. To give just one example, it has been estimated that merely proofreading the activity of the synthetases that stick amino acids on to tRNAs can cost 2–3 per cent of a cell's total energy budget. When we add in all the rest, it is small wonder that somatic cells are disposable.

I want to close this chapter by making a point which is so obvious that it hardly needs making – especially when we consider the totality of what goes on in our cells, as we have done here – but which is stubbornly overlooked time and time again: *there is probably no single mechanism of ageing.*

Leo Szilard, the nuclear physicist we encountered at the start of this chapter, suggested that mutations might be the cause of ageing. He and others gave us the somatic mutation theory of ageing. In 1956, Denham Harman gave us the free radical theory of ageing. In 1963, Leslie Orgel gave us the error catastrophe theory of ageing. And the list goes on. Zhores Medvedev some years ago compiled a list of nearly 300 different theories of varying kinds. Evidence in support of most of the major theories has been amassed, but none of them is able to explain all of the data.

If the disposable soma theory is right, each of the mechanisms that suggests an accumulation of damage in the somatic cells of the body is likely to be a part of the picture, but only a part. We need to build this concept firmly into our thinking about ageing, and we will find that it helps us to integrate the multifaceted nature of the problem rather well.

Axel Kowald, a biochemist at the Humboldt University in Berlin, and I have developed what we call a 'network theory' that makes an encouraging start in this direction. We modelled the interactions of free radicals, antioxidants, faulty proteins, scavenging enzymes and mitochondrial DNA mutations, and showed that the combined model could explain much more of the data than any single process could do on its own. It also allowed us to study the synergy and interaction between different mechanisms.

What we found was that the different possible causes of cellular ageing could all contribute to the eventual breakdown and death of the cell. However, depending on the kind of cell in the body one looks at – for example, a dividing cell like a fibroblast versus a post-mitotic cell like a neurone – and the stage in the breakdown process at which one makes the observation, these multiple causes can assume different relative importance. We need this kind of powerful analytic tool if we are to unravel the deeper complexities of how molecular mistakes contribute to the ageing process.

What is true of diverse mechanisms within individual cells is also true of the various organs and tissues of the body. In the next chapter, we will look at this aspect of how we age.

CHAPTER 9

Organs and orchestras

O, how shall summer's honey breath hold out
Against the wrackful siege of batt'ring days
When rocks impregnable are not so stout,
Nor gates of steel so strong, but Time decays?

William Shakespeare, 'Sonnet No. 65'.

One of the quirks of human mortality is that we live our lives knowing we must die some day, but we do not, as a rule, know what we will die of. Some years ago, the rattle of the letter box in my front door announced the arrival of some mail. The mail turned out to be the newsletter of the local Anglican church. On the front page was a short article by the vicar's wife, who was dying of cancer. This valiant woman, who for many years ran the local pre-school playgroup and was a favourite of generations of children, including my own, wrote that after years of vague anxiety about death and dying, she found strange comfort in knowing, at last, how and approximately when she would die.

I was shocked and saddened on reading this article, for I did not know that the vicar's wife was ill, and I have to confess that my first reaction was to disbelieve the idea that knowledge of coming death could be any kind of comfort. The British magazine *Punch* in 1895 published a cartoon showing a nervous curate breakfasting with his bishop above the following caption: 'Bishop: I'm afraid you've got the bad egg, Mr Jones. Curate: Oh no, My Lord, I assure you! Parts of it are excellent!'

I was inclined to think that the same philosophy of desperately finding good in an appalling situation was behind the article. But until it happens, one cannot really know one's reaction to the diagnosis of terminal illness. My heart tells me now that the vicar's wife was being characteristically, if surprisingly, honest.

The fact that our death is inevitable and yet its cause is largely unpredictable is an odd aspect of the ageing process. Death is unpredictable because all the different organs of the body undergo deleterious change with age, and they do so at roughly the same rate.

Some age-related changes, such as the deterioration in the blood supply to the brain, causing increased risk of a stroke, have the potential to cause death. Others are non-fatal, but important to our self-image and quality of life. The wrinkling of skin and the greying of hair are two of the most obvious external signs of the ageing process, but no one dies of wrinkles or grey hair.

Not only is there a seeming co-ordination in the way our organs age, but there can also be some remarkable adaptations, in the form of adjustments to the working of the body, as we get older. The ageing of the heart provides a good example. If we measure cardiac output – that is, the rate at which the heart pumps blood around the body – in people who are free from disease and who take a normal amount of exercise, we find little change with ageing. But the heart itself undergoes considerable change. A stiffening in the walls of the major arteries results in a progressive increase in blood pressure, which imposes an increasing load on the heart. To keep the heart's output constant, the left ventricle – the chamber that pumps the blood around the body – grows bigger, altering the proportions of the heart. This compensation is not without cost. The heart has to work harder for each heart-beat and it uses more energy. The result is that the overall efficiency of the cardiovascular system drops significantly.

What we see in the ageing heart, then, are two distinctive features of the ageing process: loss of 'functional reserve' – that is, the spare capacity to cope with unusual demands – and 'adaptive adjustment' – the development of compensatory changes to help us get by. It is this capacity for adaptive adjustment, combined with the synchronicity of age changes in different organs, that gives the ageing process the appearance of being under some kind of active control.

This semblance of control, which some go so far as to call the 'wisdom' of the ageing body, has encouraged the search for a central pacemaker of the ageing process, somewhat like the conductor of a musical orchestra who calls the various instruments into play at the appropriate time and rounds the performance off in a synchronised finale. But is ageing really orchestrated, or do our organs just play along to the same rhythm, finally terminating the session when a key player has had enough? Are we orchestras of organs, or just jazz groups jamming?

In this chapter, we will look at various organs of the body as they age and ask if there is a common set of mechanisms that might explain the apparent co-ordination of the changes we observe. Of course, not all of the changes we will survey happen to everyone. In case this chapter seems too gloomy, please remember that in Chapters 15 and 16 we will look at the positive steps that can be taken to avoid them.

At the outset, let us remind ourselves that the existence of an inbuilt clock for ageing is something that we cannot take for granted. We look for clocks because we have a tendency to seek purpose in nature. But if ageing really is a purposive death mechanism, which Chapter 5 made us doubt, surely it is a little clumsy. The fact that ageing affects our non-vital organs just as much as it affects our vital organs must give further cause for doubt.

If we abandon the idea of clocks, there is another, more plausible way to explain the synchronicity of age changes which makes no call on the idea of purpose. This is to think of how the shrewd manufacturer designs disposable goods. When Henry Ford began to mass produce his motor cars, he sent teams of engineers around the scrapyards of America to find what his cars had 'died' of. Any recurring fault was referred to the designers, who were charged with correcting the weakness. The upshot was a balanced design which ensured that, on the average, each of the major components wore out at the same rate, even though the nature of the wearing-out process varied from one component to another. Ford's shrewdness as a motor car manufacturer echoed the exquisite cleverness of natural selection in designing our disposable somas. Except, of course, that natural selection was not clever at all – it just did it without thinking.

Let us begin with our teeth, because teeth epitomise the disposability of the soma. We get just two sets of teeth: the milk-teeth that we have in early childhood, and the main set that begin to emerge as we leave infancy. Some species like alligators have an endless supply of new teeth. We, like other mammals, do not.

One should not, as the saying goes, look a gift horse in the mouth. The reason one might wish to do so, of course, is that a horse's mouth, particularly the state of its teeth, gives an immediate indication of the animal's age, and hence its worth. Horses use their teeth extensively to grind their cellulose-rich food and facilitate its digestion. Over the course of a lifetime, a horse's teeth get worn down. Wild horses, if they were to live long enough, would eventually suffer malnutrition and die because of this.

Human teeth reveal age too, even though we do not use them like horses to grind grass. In the modern era with our more refined foods, teeth get less wear, but they face new challenges through our love of sugar and our fondness for hot drinks and ice cream, causing cavities and cracks. Actually, dental care is improving and a growing number of us now keep at least some of our teeth throughout life. Gone are the days, in developed countries at least, when parents gave as wedding gifts to their daughters a full dental extraction and set of false teeth. But old age still shows in the mouth because, even if the teeth themselves remain sound, the slow shrinkage of the jaw bones leaves the teeth less well anchored, and the receding of gum tissue, caused as much by repeated gingival infection as by intrinsic ageing, leaves dentine exposed.

Bones are the second port of call in our tour around the ageing body. Bones are very suitable organs for the student of ageing because, as any 4-year-old can tell you, our skeletons are all that remain of our earthly bodies when we are dead and gone (unless, of course, we are cremated). But the bones in a living body are not the dead things of graveyards and museums. They are living structures with cells and blood supplies, which grow with us through childhood and decline with us in old age. The ageing of bones is one of the most important causes of loss of mobility in older people, especially women.

One-third of women over the age of 65 in the United States will have collapsed spinal vertebrae as a result of bone thinning, or osteoporosis. By age 90, one in three women and one in six men will have suffered a fracture of the hip. This devastating fracture, and the stress to the system that results from it, often triggers a downward spiral that results in death. Of those who survive, as many as half may suffer permanently impaired mobility and be unable to walk again without assistance.

Bones in old people become more liable to fracture, even after a minor stress, because they become thinner and lose their mechanical strength. The strength of a bone depends on how dense it was in the first place, during young adulthood, and on how fast it has thinned during the process of ageing. There are significant differences between races and the sexes in the weight of the skeleton when it is at its peak. The native inhabitants of Navrongo, like other West Africans, have high bone density compared with white visitors from Europe. These racial differences persist among African-Americans and Americans of European descent. The more bone you have when you are a young adult, the less likely you are to develop osteoporosis.

The loss of bone is a universal process in the general population due to the decreased function of the bone cells that maintain and remodel the skeleton, and due to the impaired absorption of calcium from the diet. Bone is continually worked on by the combined actions of two kinds of cell, called osteoclasts and osteoblasts. The osteoclasts nibble away at the surface of the bone, forming small indentations or grooves. The osteoblasts then come along and lay down new bone mineral to fill in the holes made by the osteoclasts. The osteoclasts and osteoblasts operate together, like teams of workmen stripping and relaying the surfaces of roadways to keep them in good shape. The reason is probably the same: to strip away old surface bone with its microdamage and cracks, and to replace it with new.

With ageing, what seems to happen is that the osteoblasts fail to keep pace with the osteoclasts, perhaps because the coupling of the actions of the two kinds of cell becomes a little impaired. Imagine a poorly supervised road gang where the resurfacers fail to notice all of the sections where the surface has been stripped. The upshot will be that gradually the road surface gets worse and worse, and eventually it breaks up entirely.

Within the skeleton are two major types of bone: cortical bone, which is the continuous material that makes up the long bones of the arms and legs, and so on; and trabecular bone, which is a fine meshwork found inside the vertebrae and inside the ends of the hip bones. Both kinds of bone lose mass during ageing, but the effect on trabecular bone can be particularly marked. This is because the individual elements in the lattice of trabecular bone are extremely thin even in a young person, so that when they get thinner as a result of bone loss during ageing some of them will disappear altogether. The result is a pronounced loss in mechanical strength and is the reason why osteoporotic vertebrae collapse in upon themselves, producing a reduction in stature and spinal curvature. Spinal curvature results in the so-called dowager's hump and in extreme cases condemns sufferers to spending the remainder of their lives with faces turned permanently to their feet.

Women are particularly afflicted with osteoporosis because, in addition to the gradual loss of bone through general bone ageing, women suffer an accelerated bone loss following menopause, which results from the decline in sex hormones like oestrogen. Bone loss also occurs during the period that a mother breastfeeds her children. Although this is normally a temporary loss that gets made up after weaning, some permanent loss can occur, especially in women whose diet is deficient in calcium. The accelerated phase of bone loss after menopause extends for about 10 years and can destroy an extra 10–15 per cent of cortical bone and 15–20 per cent of trabecular bone. The post-menopausal bone loss in women can be substantially reduced by hormone replacement therapy, which is particularly valuable for women with below average bone density. However, the general age-related decline in bone mass cannot yet be prevented or appreciably slowed.

Tooth decay and osteoporosis leave permanent records in the skeleton and so too do two of the commonest complaints of old age: rheumatism and arthritis. When the mummified remains of a man who died about 5,000 years ago were found recently at the Hauslabjoch glacier in the Otztal Alps, pathologists could determine that the man may have been 50 or more because of the evidence of osteoarthritis in his joints. Many a life has been made miserable in old age by the pain and disability inflicted by rheumatism and arthritis.

Rheumatism and arthritis are terms that describe not single disorders, but ranges of conditions, the precise diagnosis of which can be far from straightforward. Pain is felt mainly when flexing the muscles and joints, which can become severely swollen and deformed. When an X-ray is taken, some abnormality of the joints can be seen in nearly all people aged 65 years or more. These changes are classed as moderate or severe in nearly two-thirds of cases, half of whom have three or more joints affected. The good news is that not all of the individuals whose X-rays reveal abnormalities have symptoms. Even so, the scale of suffering is considerable.

The major rheumatic and arthritic conditions affecting older people are rheumatoid arthritis, osteoarthritis, polymyalgia rheumatica and gout. Rheumatoid arthritis is a condition involving inflammation of the joints and can lead to serious deformation and immobility. Rheumatoid arthritis is by no means confined to the elderly, and indeed it is commonly held that the most frequent age of onset is in mid-life, between the ages of 35 and 55. However, there is evidence that a considerable number of cases among the elderly never come to the attention of a rheumatologist and get labelled as some other inflammatory condition.

In old people, as in young, the onset of rheumatoid arthritis is most often gradual, involving a stiffening and swelling of the joints of the fingers, hands, wrists and knees over a number of months, accompanied by some aching and restricted movement of the neck, shoulders and hips. Sometimes, however, the condition develops more suddenly within weeks, or even days, and in these cases the inflammation of the joints is accompanied by fever. The disease is autoimmune by nature – that is, it involves attack by the immune system against the body's own tissues – and at some course in its progression it is associated with the appearance in the blood of a protein known as rheumatoid factor.

The causes of rheumatoid arthritis are as yet unknown, but it is likely that in at least some cases the trouble is triggered by infection by a bacterium or a virus. The role of infective agents in triggering autoimmune diseases has proved hard to pin down because the troublesome bug does not need to linger to do damage. It starts an abnormal reaction that continues after it is gone – a classic hit-and-run accident.

Like other conditions that involve the destruction of body tissues by faulty immune system reactions – multiple sclerosis is another example – rheumatoid arthritis can go through active stages, when the disease progresses, new arthritic nodules form, and further destruction of the joint tissue occurs, or it can pass into a relatively inactive stage, when the sufferer has only to cope with the physical limitations of the damage that has already occurred. Either way, rheumatoid arthritis can be a real pain.

Polymyalgia rheumatica is a disorder that is more strongly age-associated than rheumatoid arthritis, being rare at younger ages and becoming much commoner in old age. Polymyalgia rheumatica is often confused with rheumatoid arthritis because it too can affect the joints, the knees and wrists being the usual targets. The classic hallmarks of polymyalgia rheumatica are pain and stiffness in the neck, shoulders, lower back and pelvic girdle, which in some cases comes on quite suddenly, but in other cases takes weeks or even months to develop. Untreated, the condition persists, but does not usually progress. However, it is a good idea to check if a patient has polymyalgia rheumatica, as opposed to other rheumatic conditions, because it responds well to low doses of corticosteroids and there is a small risk that it may progress to a more serious condition called arteritis. Arteritis is an inflammation of the blood vessels, which can cause severe headache and damage to the nervous system, even triggering blood clots that might, for example, cause a stroke.

Gout is a term that conjures up images of florid gentlemen in bath chairs who are suffering the consequences of a lifetime of over-indulgence in rich foods and port wine. 'Gout', wrote Lord Chesterfield in the eighteenth century, 'is the distemper of a gentleman'. This is unfair to many of its sufferers. While lifestyle does appear to play a part in some cases, gout is often the result of an inherited abnormality in metabolism that leads, over the years, to the formation of small crystals of a substance called sodium urate in the cartilage of the joints. These crystals cause extreme sensitivity and pain. When gout strikes early in middle age it targets men more than women, but at older ages women are affected nearly as often as men and the cause is commonly associated with some other problem, such as a haematological disorder.

Osteoarthritis, a breakdown in the working surfaces of the joints resulting in pain and lost flexibility, is the commonest of the arthritic conditions that affect older people, and it is the one most often associated with wear and tear. The knee and hip are the joints that cause the greatest trouble. Just why osteoarthritis occurs is not known for certain, and there are many theories, most of which probably contain some truth.

The wear-and-tear nature of osteoarthritis – that is, the accumulation of damage throughout life – is supported by the fact that certain occupations, such as ballet dancing and operating a pneumatic drill, carry increased risks. There are not many shared features in the professional lives of ballerinas and pneumatic drill operators, but wear on the joints is one. Obese people also have increased risks of osteoarthritis, which is likely to arise from increased load bearing on the surfaces of the knee and hip joints.

In addition to mechanical damage, the causes of osteoarthritis are thought to involve a gradual deterioration with age in the ability of cells to manufacture the special molecules that make up the working surfaces of the joints, and to receive and correctly interpret the various signals, internal and external, to which these cells must respond. It is easy to see how this kind of generalised breakdown can exacerbate the problems caused by mechanical damage. Imagine a motor repair workshop where the engineers are gradually ageing. Their ability to keep pace with a constant stream of breakdowns and accidents will slowly be impaired. Now if the accident rate were to increase because the drivers are ageing too, the situation would eventually become unworkable.

After rheumatism and arthritis, the commonest complaint of older people is that they can no longer remember as well as they could when they were younger. Inability to remember may cause frustration, embarrassment and a sense of diminished identity. However, the scale of memory loss tends to be exaggerated by our unfortunate habit of attributing any small act of forgetfulness to the effects of ageing, even if neither the forgetter is very old, nor the memory lapse any different from the kind that is also common in youth. Nevertheless, memory does genuinely suffer from ageing.

Measuring memory is a tricky business because there is a large subjective element in how we rate our own ability to recall past events. If our memory does start to decline, we might even forget what we have forgotten! Fortunately, a large body of psychological research has been devoted to the study of memory and its alteration with age.

One of the enduring legends about memory in old age is the

notion that old people can recall their early lives better than they remember the events of last week. So widely held is this belief that it even has a name – Ribot's law – after a nineteenth-century psychologist who suggested that, as memory degrades in old age, those memories that were laid down first gradually get uncovered as the later memories get stripped away. Picture a whirlwind passing through the office of an untidy professor, blowing away the papers on top of the dusty piles and revealing the early works that lie beneath. This is Ribot's law.

Actually, Ribot's law is a myth. We must accept at face value the oft-repeated claim that early memories are particularly vivid. Only those who are doing the remembering can assess how vividly the images come back. But researchers who have been to great pains to check the accuracy of those early memories which can be verified have found that the recollection of these early events is not especially accurate.

One of the things each of us who was alive at the time is supposed to remember is what we were doing when we heard the news of President John F. Kennedy's assassination. I remember it as if it were yesterday, how my dancing class at junior school was sent home early, and how I broke the news to my family. The trouble is, none of my family's recollections tallies with my own, or with each other's!

Patrick Rabbitt, director of the Age and Cognitive Performance Research Centre in the University of Manchester, has shown that one of the main factors linked to the vividness of memories is rehearsal, the repetition of the remembered item many times for oneself, one's friends and one's family. The early years of life tend to be particularly rich in significant events that are more emotionally evocative and tend to be rehearsed most often.

Before we dig deeper into the complexity of how memory alters with age, let us pause to ask why memory exists at all. What is the function of memory, and how does it work? If we had no memory, we would have no awareness of time. We would be locked in an eternal now, with no ability to draw on experience and no ability to foresee the future. Not only would we not do things like remember to buy toothpaste the next time we go shopping, but we could not even brush our teeth. An action like tooth brushing requires that you hold in memory what a toothbrush is, what you use it for, and which teeth you have brushed already. Take that away and you would simply stand, toothbrush in mouth, until the next impulsive action struck you. More significantly, an absence of memory would make speech an impossibility.

We can see the importance of memory when we think about how a computer functions. A computer has a microchip called a processor, which is where all operations in the binary language of computer-speak take place. But a processor is no use if it is not accompanied by memory, because memory is where the data bits reside before they are processed and where the output bits go when the process has been performed. Programs and data in current use are held in an area of working memory called RAM (random access memory). Bits of information shuttle in and out of this 'working memory' at great speed. Long-term storage is managed on larger data stores, the disks, from which recall takes more time. The speed of a computer's 'thought' is a function both of the basic operating speed of the processor (how fast the data bits are dealt with) and the size of the RAM (a smaller RAM means more delays, as information gets swapped back and forth between working memory and the disk).

Human memory appears to be a little bit like computer memory, in that we seem to have different levels of data storage and different kinds of recall, but the analogy cannot be pushed very far. The human brain relies on a huge network of units, the neurones, for its thought processes, and is not as fastidious about accuracy as a computer. All of us who use computers know how frustratingly precise a computer can be if it requires a zero instead of a capital O to recognise a command. On the other hand, we would pretty soon get fed up if a computer remembered only vaguely the letter we typed into our word-processor last week. The human brain is exquisitely good at selective attention (the ability to ignore the irrelevant), fuzzy logic (the capacity to take short cuts) and filtering memories. The filter does not always work perfectly, but perfect memory would be more of a curse than a blessing. It is the sort of thing you might wish for from the genie in the lamp, but would regret the minute the genie had gone. Just think how awful it would be if we remembered every unselected detail of our past lives. 'And we forget because we must, and not because we will', wrote Matthew Arnold in his poem 'Absence' in 1852.

It has long been known that ageing is accompanied by a general trend towards poorer memory performance and the slowing of mental processing. But like any ageing trend, there is great variability among individuals. Those whose memory declines the fastest are often found to have other problems, such as cardiovascular disease or diabetes.

What causes memory to decline with age? One of the most commonly held beliefs about cognitive ageing is that brain function declines because our neurones die off. This is only part of the story. It is true, of course, that neuronal loss is a one-way street; no new neurones are formed to replace any that do die. But it now appears that the scale of neuronal loss in the healthy aged brain has been exaggerated. The brain as a whole shrinks, but there are other cells that make up the mass of the brain as well as neurones. These provide the supporting framework in which the neurones are housed and much of the brain shrinkage is due to losses of these cells. The other factor that may explain a good many features of normal brain ageing is that the processing of information slows down. Because brain cells do not divide, they have to live with their metabolic waste products. A dividing cell is always diluting any junk with the new cell material that gets formed between divisions. Half the junk goes to each daughter, which is a good way of keeping the molecular clutter under control. Brain cells have to dispose of their own waste, but some is hard to get rid of. There is good evidence that a slow build-up occurs in non-dividing cells of molecular waste products that are resistant to removal by molecular garbage disposal. It may be this junk that impedes the functions of older neurones and slows them down.

A simple reduction in the processing speed of the neurones can have all sorts of effects on memory and cognitive performance. Older people have been shown to remember fewer items from lists than younger people, but they can often remember as much – and as accurately – as the young if they are given longer learning times. It has also been shown that age slows difficult decision making. Older people who have kept diaries throughout their lives tend to use shorter sentences, with less grammatical complexity, than they did in middle age. 'Tip of the tongue' experiences, when a word or name eludes recall, become more frequent and the lost item takes longer to surface.

An aspect of memory that also alters with age is the ability to suppress irrelevant information that comes to mind. When given a problem to solve where only some of the information is relevant, age brings increasing difficulty in separating the relevant from the irrelevant. This may be why in social contexts older people are more likely to digress from a topic and less able to remember who said what.

Small wonder that memory loss and cognitive impairment are such frequent complaints about the frustrations of old age. But there are memory-enhancing tricks that work at least as well for old people as they do for young, and there is good evidence that regular practice of mental skills – through activities like chess, bridge and crossword puzzle solving – can help.

Worse than memory decline, although arguably less of a problem for the affected individual than for family and carers, is dementia. How many of us, when faced with a sufferer from late-stage Alzheimer's disease, secretly wish that we might die before we become demented?

The growth in the number of people with dementia is directly related to increased life expectancy. At age 65, less than one in a hundred of the population in the United States and Europe suffers from dementia. By age 85, this figure has climbed to one in six. Alzheimer's disease is the leading cause.

The hallmarks of Alzheimer's disease are a wasting of certain parts of the brain and the presence within brain tissue of two kinds of small lesion called neuritic plaques and neurofibrillary tangles. The plaques and tangles can currently be measured only at autopsy, so during life the clinical diagnosis of Alzheimer's disease is somewhat tentative.

What those who regularly examine the diseased brain tissue describe is that the wasting occurs mainly (but not exclusively) in the outer surface of the cerebral cortex – the convoluted outer layer of the brain – and involves the largest of nerve cells. This apparently specific loss is mirrored in a reduced level of chemicals called neurotransmitters, which these nerve cells produce as a part of their normal function. Altogether it is estimated that less than 15 per cent of the nerve cells in the cerebral cortex are affected in Alzheimer's disease patients, but it may be that these cells play a particular role in awareness, memory and reasoning. It is a feature of the disease that some patients remain in good health apart from the devastating changes that rob them of so much of their identity. The destruction of the cells that really matter can be cruelly selective.

The neuritic plaques seen in Alzheimer's disease consist of clusters of degenerating nerve cell endings surrounding an amorphous core. The core is made up of fibres of beta-amyloid, an abnormal fragment of a bigger protein. Beta-amyloid is a stubborn molecule that resists the usual scavenging mechanisms which clear away unwanted proteins. The neurofibrillary tangles are nerve cells that are filled with another kind of abnormal protein fibres, called paired helical filaments. Like beta-amyloid, these filaments are aberrant forms of normal proteins and are peculiarly resistant to removal by protein scavengers.

Just what causes the plaques and tangles to form, and if and how these actually cause dementia, is not yet known. Scientists still have to establish for certain whether these tell-tale signs are the trigger of what goes wrong, or just the wreckage that some other damaging process leaves behind. There are some who believe that periodic inflammation of parts of the brain is the real culprit, and a significant development that lends support to this hypothesis has been the discovery that regular use of certain anti-inflammatory drugs, like the common pain-killer ibuprofen, may delay the onset of Alzheimer's disease.

One of the more puzzling things is that many perfectly normal old people show plaques and tangles within their brains without being demented. This suggests that, in spite of the change in status when Alzheimer's disease was labelled as something different from plain old age, it is part of a continuum after all. This holds out the hope that all the research effort that has gone into Alzheimer's disease in recent decades will help us to understand 'normal' brain ageing as well.

Genetics is providing some important clues. The gene for the amyloid protein is found on human chromosome 21, the same chromosome that is present as an extra copy in Down's syndrome. People with Down's syndrome typically develop brain alterations just like Alzheimer's disease if they live beyond age 40. There are families in which there is a high inherited risk of developing Alzheimer's disease, and in which the onset of the disease tends to occur much earlier than in the general population, sometimes as early as in the forties. In some of these families, the genetic fault has been found to occur within the gene that produces the amyloid protein, from which the abnormal beta-amyloid derives. But there are families with inherited Alzheimer's disease where genes on other chromosomes are involved, and there are certain common genes within the general population that have also been identified as risk factors. We will look further at the complicated genetics of Alzheimer's disease in Chapter 14.

Even in families with inherited risk of Alzheimer's disease, it is not automatic that the disease will develop. What appears to be inherited is merely an increased risk, and not a certainty. It may be that some other co-factor is necessary. For many years it has been thought that perhaps aluminium is such a co-factor, since in Alzheimer's disease there is an accumulation of aluminium, especially in the areas of the cerebral cortex with many neurofibrillary tangles. However, attempts to prove a link between consumption of excess aluminium and the risk of Alzheimer's disease have generally failed, so it may be that the accumulation of aluminium is a consequence and not a cause of the disease.

Another possible co-factor is a virus, specifically the cold-sore virus, herpes simplex. Recent epidemiological findings suggest an association between the presence of herpes simplex virus and Alzheimer's disease. The researchers found that when they compared brain tissue from Alzheimer's disease patients and non-Alzheimer controls, the Alzheimer brain samples were more likely to test positive for herpes simplex virus, especially if they came from patients who carried a particular variant (the epsilon 4 allele) of the gene called apolipoprotein E.

Given the high profile of Alzheimer's disease nowadays, one might easily gain the impression that this is the only kind of dementia affecting older people. This would be wrong. As many as one in five of people hospitalised for dementia suffer from Lewy body dementia, so called because the disease is characterised by the presence of abnormal structures called Lewy bodies in the neurones of the cerebral cortex. Lewy body dementia follows a more rapid course than Alzheimer's disease, with a tendency to visual hallucinations and transient loss of consciousness. Lewy bodies are also involved in another scourge of later life, Parkinson's disease.

In Parkinson's disease the damage occurs not in the cortex, but in a part of the brain called the substantia nigra. The neurones of the substantia nigra issue signals controlling movement, so the loss of these cells is why Parkinson's sufferers exhibit a worsening tremor in their movements, loss of balance, and have more and more difficulty initiating actions like rising from a chair. It is thought that oxidative damage by free radicals plays an important role in causing Parkinson's disease and perhaps also in Lewy body dementia.

Everyone knows that blood flow to the brain is crucial. Stop it even for a short while and brain death occurs. But our brains can suffer trouble with their blood supply that falls short of total brain death, yet destroys brain tissue just the same. With advancing age, the small blood vessels of the brain can become choked and an insidious kind of dementia, vascular dementia, can result. A more major interruption to the brain's blood supply causes a stroke.

Strokes happen when the loss of blood supply destroys a whole section of brain tissue. Like heart attacks, strokes can occur in young people, but they become much commoner with age. A stroke can affect almost any aspect of brain function, and can have widespread or localised effects. It is hard to say whether it is worse to lose awareness and reasoning gradually over a wide area of brain function, or to lose the power of sight, speech or movement, while remaining fully alert.

A stroke can be mild or devastating, depending on which part and how large a volume of the brain is affected. After the first wave of damage caused directly by the loss of blood supply, there is often a secondary wave of damage caused by cell suicide in areas around the primary lesion, even if the blood supply is restored. Just why this happens is not yet understood and it may simply be that the first cells to die send out a chemical alarm signal that inadvertently triggers the surrounding cells to destroy themselves. As we saw in Chapter 7, all cells have the potential to commit suicide, and on a small scale this is useful in that it provides a tidy way to get rid of a damaged cell without the risk of starting an inflammatory reaction. There is no obvious logic in widespread cell suicide, however, and if we can find ways to block the suicide response that is responsible for the secondary wave of cell death, it may become possible to lessen the damage caused by strokes.

Loss of blood supply in the brain is but one aspect of a much bigger problem of old age: the degeneration of the heart and blood vessels throughout the body. The organisation of the blood supply is a masterpiece of evolution. The blood, carrying its vital cargo of oxygen and nutrients, must reach the remotest cells of the body and return, carrying waste products, to be cleansed through the liver and kidneys before being pumped once more to the lungs to become freshly oxygenated again.

Blood vessels, particularly the arteries, require very special properties of elasticity and strength in order to cope with a lifetime of stretching and contracting as the heart pumps the blood at pressure around the meshwork of big, medium and tiny tubes that together make up the circulatory system. As the blood flows along a blood vessel, the drag of the fluid against the vessel wall exerts what is known as a 'shear' force, similar to the force that causes a river eventually to erode a large valley. The shear forces on the vessel walls are relentless agents of wear and tear. The harmful effects of these shear forces show up in the deadly process known as atherosclerosis.

Atherosclerosis begins surprisingly early in life and develops progressively over the years. Early signs can be detected at autopsy even in fit young adults, no more than 20 years old, who have died in accidents. Atherosclerosis consists of the formation of lesions in an artery wall, which become the focus for the formation of blood clots. Typically, the process begins when a region of the endothelial cells lining the artery becomes damaged. The damage may be minor, but it can result in the endothelial cells temporarily failing to hold securely to each other and to the underlying connective tissue. Because the lining of the blood vessel is continually subject to the shearing action of the blood flowing past, a patch of the endothelial cells may get stripped away, exposing lower layers of the wall. Now a repair is needed.

When a plumber repairs a damaged pipe, he can usually shut off the water flow in order to effect the repair. This luxury is not available to your arteries. The exposure of lower layers of tissue in the artery wall encourages the sticking of platelets in the blood. Platelets are small capsules packed with powerful clotting enzymes, whose prime function is to guard against bleeding when injury occurs. Once the first platelets begin to stick, others stick to them to form a clump. The platelets also release a substance called platelet-derived growth factor, or PDGF. PDGF is a strong stimulant that causes cells to divide, and its action can stimulate cells from the lower levels of the artery wall to grow into the site of the original injury, creating an interruption to the smooth endothelial wall, much like a patch on the inner tube of a bicycle tyre.

The result is that repeated damage to the artery wall can create lesions that grow progressively over the years, until finally they give rise to the classic hallmarks of cardiovascular disease, restricting blood flow, increasing the risk of a serious clot and maybe triggering a heart attack, stroke or pulmonary embolism (a clot that blocks blood supply to the lungs).

In general, as the cardiovascular system ages, its tissues lose some of their elasticity and stiffen, mainly because of subtle changes in the matrix of proteins and other materials outside the cells that are the result of all kinds of random molecular mishaps. Proteins get oxidised or modified in other ways, and molecules form spontaneous chemical cross-links that reduce flexibility. The process is not unlike the gradual stiffening of a rubber band, which when exposed to light and air eventually becomes brittle, and breaks. Loss of elasticity occurs in nearly all of the tissues of the body as we age, but it is particularly a threat in tissues that need to flex in order to work properly. Top of the list of such tissues are our blood vessels and our lungs.

Ageing of the lungs is a serious source of disablement and, ultimately, death. In youth, a healthy pair of lungs has a huge reserve capacity – even world-class athletes rarely push their lungs to the limit. But age-related loss of elasticity in the intricate network of sacks and chambers that hold the air we breath, means that when we are old we can neither fill our lungs as full, nor empty them as completely of stale air, as we could when we were younger.

Our lungs are another masterpiece of evolution and the fluid that our airways produce to effect the correct gas exchange is chemically very special. Fortunately, there appears to be no major change in gas-exchange efficiency with age. However, with ageing comes a marked increase in emphysema, a pernicious lung disease that involves a build-up of fluid inside the air spaces. There is also an increase in bronchitis, causing constriction of the airways. For large numbers of old people, life becomes a continual fight for breath that grows increasingly short.

The airways are also an important interface between our bodies and our environments, with airborne pathogens forever seeking a way in. Thanks to vaccines and antibiotics, respiratory infections are no longer the huge killers that they were in the nineteenth century, although in the area around Navrongo they still cause many preventable deaths. But as our lungs age, we are increasingly prone to pick up infections, and these infections often take longer to come to medical attention because the fever response in old people works less well. Among the elderly, even in the developed countries, pneumonia remains a frequent cause of death.

The battle against infection is fought at many levels in the body and the front-line troops in most of these skirmishes are the cells and antibodies of the immune system. The immune system is extremely complex. As they have grappled over the years with the intricacies of the immune system, immunologists have needed to develop an arcane terminology to describe all of the various parts and how they interact. We will skip most of the terminology here in order to get to the essence of what goes on.

The central problem is that the immune system must protect our bodies against an array of unknown and unpredictable assailants, so it builds up a repertoire of weapons of quite extraordinary versatility. It does this by means of a remarkable set of processes. The key is the production of a hugely diverse set of white blood cells, or lymphocytes. The diversity is generated by having a part of the gene sequence within the lymphocytes undergo mutation at an unusually high rate. The mutated sequences make proteins that form a part of the cell surface, and which take part in the recognition and destruction of foreign invading pathogens. Because the mutations throw up such an extraordinary variety of forms, it is likely that among them there will be at least one that will recognise a part of each intruder's molecular structure, even if this particular intruder has never been seen before.

So far, so good. However, we come up immediately against two big problems. Firstly, it is fine that the immune system can produce all of this diversity, but if a new virus comes along and only one in a billion lymphocytes can recognise it, what use is that? The immune system has to react by getting the one in a billion cells to undergo a rapid series of cell divisions and make many copies of itself. This is called clonal expansion. And when the virus has been cleared, the immune system remembers its visit. Although most of the expanded clones fade away in the normal course of cell turnover and replacement, the immune system keeps a guard of so-called memory cells at the ready, so that if the same invader should show its face again, the immune system can respond more quickly the next time. This is the basis of vaccination. In time, the naive immune system of the newborn infant matures into the street-wise immune system of the adolescent, which has seen it all before. Even so, the immune system keeps one eye open for fresh trouble.

The other big problem is that many of the lymphocytes created by the mutation mechanism recognise the body's own cells. These cannot be allowed to proliferate or they would cause an autoimmune reaction, leading to the destruction of healthy tissue by 'friendly fire'. So the immune system marshals all of its new recruits through a remorseless finishing school, where any that react to 'self' are made to commit suicide. This is called clonal deletion.

There are in fact separate finishing schools for the two major kinds of lymphocyte, the T-cells and the B-cells. The B-cells (antibody-producing cells) go through their screening process entirely within the bone marrow, where they are formed. The T-cells (which have a rather more complicated repertoire of functions) are also formed in the bone marrow, but they mature and get screened in an organ called the thymus, which is located at the base of the neck.* The thymus has a rather special place in the history of ageing 'clocks' because for many years it was believed that the thymus was the pacemaker of ageing. The reason for this quaint idea was simply that it is one of the few organs of the body that starts to get smaller quite early in life. It reaches its maximum size within our first year, and then it reduces gradually and progressively throughout life.

Just why anyone should think that a shrinking thymus would be the pacemaker of ageing has always puzzled me. Mary Ritter, professor of immunology at the Royal Postgraduate Medical School in London, has a more prosaic but satisfactory explanation. As our immune system matures, Ritter suggests, the need to push new recruits through the finishing school declines. The process of making, testing and deleting so many candidate T-cells is metabolically expensive and it makes sense to build the thymus in such a manner that it gradually fades away.

The shrinking of the thymus has little direct effect on immune functions in old age. The ageing of the immune system is more subtle. We see a general impairment in many aspects of immune function, but also great variability between individuals. The variability is probably a reflection not only of genetic differences, but also of the fact that each of us has a different history of infectious disease, and so we tax our immune systems in different ways. We saw in Chapter 7 how there is some suggestion that the T-cells in AIDS patients eventually run out of proliferative capacity, and such a process could operate more generally as part of the ageing of the immune system. But the massive and remorseless destruction of the particular subset of T-cells that get infected by the human immunodeficiency virus (HIV) is a good deal more extreme than normally occurs, and most of us appear to live out our lives without exhausting the capacity of the immune system to produce the necessary cells.

* If you have ever eaten calf or lamb sweatbreads, you have eaten a thymus.

What perhaps better explains the impairment in immune function in old age is simply that the tightly regulated network of cell-to-cell interactions begins to get a tiny bit sloppier. In a set-up as highly interactive as the immune system, a little noise can create appreciable disruption. The other thing that happens is that we see an age-related increase in autoimmune reactions. Once again, this probably reflects the beginnings of a breakdown in the cell recognition and signalling systems that so stringently delete the potential troublemakers as they pass through their finishing schools.

The immune system represents one kind of interaction between our bodies and our environment. Another occurs through our senses. Sight, hearing, taste, smell and touch are all affected by age to some degree. Many of us regard the loss of sight and hearing as two of the greatest dreads of old age.

Ageing affects sight in several ways. The eye lens becomes clouded through cataract; the cells of the optic nerve may be damaged by the effects of glaucoma (raised pressure within the eyeball); or the cells of the retina that react to the light image falling upon them may become damaged or lost in a condition called age-related macular degeneration, or ARMD.

The eye lens is not, of course, an inert material like glass or plastic, but is made up of some very remarkable cells, which have evolved to be almost completely translucent. To achieve this extraordinary feat, the lens cells have had to forgo having things like nuclei or mitochondria inside them, because these would obstruct the passage of light. Without a nucleus, a cell has no blueprint for making proteins. This means that the cells of your eye lens made their major protein components very early in life, while you were still in your mother's womb. These same proteins have to give a lifetime's service. We saw in Chapter 8 how vulnerable proteins are to all sorts of damage, and it turns out that lens proteins have special mechanisms to protect them. There are potent scavengers of free radicals in lens cells, which get topped up from neighbouring cells in the eye. And lens proteins are designed to be particularly resistant to many of the kinds of chemical reaction that might cause damage. Nevertheless, over the years the proteins of the lens gradually suffer reactions caused by heat, oxidation and interactions with sugar molecules. More and more of the light that falls upon the lens gets scattered instead of passing through, and in time the lens grows opaque. This is how a cataract forms.

The business end of the eye is the retina, where the inverted image of the outside world falls, just like the image from a pin-hole camera. To translate the light image on the retina into the electrical pulses that get conducted along the optic nerve to the brain, the cells of the retina perform a unique set of chemical reactions. The photons of light that fall on the visual pigments of the rod and cone receptor cells get absorbed, and yield up their energy. This is what gets the special chemistry going. But light is also damaging, and in fact light as well as oxygen can trigger the formation of free radicals. The cells of the retina are geared up to deal with the by-products of all this light-activated chemistry, but it appears that over time damage inevitably accumulates. Some of this damage takes the form of a build-up of a kind of molecular sludge called lipofuscin. Gradually the fine balance of the retinal chemistry gets disturbed, and this leads to cellular changes in the sensitive membranes of the retina. Fluid leakage and haemorrhage across the membranes can cause a catastrophic loss of vision in the central part of the visual field, the macula, which is where the burden of light-activated chemistry is greatest.

When I was in junior school, I had a Kodak Box Brownie camera. Unnoticed by me, the clasp of the camera case rubbed directly against the camera lens and in time the centre of the lens got badly scratched and worn. The result was that my photos came to be increasingly fuzzy and eventually useless in the middle. The effects of ARMD are like this.

The ageing ear is a little microcosm of the ageing body because the mechanics of hearing involves skin, bones, joints, muscles, tissue elasticity, nerves and blood supply. Sound is channelled from the outside world along the tunnel of the outer ear, and it fetches up against the eardrum where it causes vibration. This vibration is transmitted to the bones of the middle ear. The vibration in the bones of the middle ear is finally transmitted to the fluid of the inner ear, where it stimulates nerve cells that connect with the auditory centres of the brain. Any part of this delicate machinery can go wrong with age, and the effects of age upon hearing are extremely variable. Most hearing loss results from degeneration of the inner ear affecting the remarkable organ of Corti, where the crucial 'hair' cells that turn sound vibration into nerve impulses are gradually lost. However, a number of the hearing problems experienced by older people are actually caused by skin changes in the outer ear channel. Sweat glands in the skin die off one by one, the ear wax becomes drier, and hard wax simply builds up to block out sounds.

There is, of course, much more to the ageing of skin than its effects on hearing. Skin is the most visible organ of the body and when examined closely it has probably the most complicated cellular architecture, with its multiple layers punctuated with hair follicles, sweat ducts and sebaceous glands. All of the tissues of the skin undergo some change with age, wrinkles and the greying of hair being the best known. It is clear, however, that the skin has a considerable functional reserve, because although the skin loses elasticity and heals wounds more slowly with advancing age, it never really wears out. A considerable fraction of the age-related lesions of the skin can be identified with damage from agents such as sunlight. The appearance of the skin is intimately connected with our perception of ageing, and to have visibly old skin has a significant emotional impact on us.

We have toured many of the organs of the body and found many, diverse changes that occur with age, but we have not so far found evidence of a conductor to co-ordinate and direct these changes. Perhaps the strongest candidate for such a role is a system that we have not yet considered: namely, the glandular, or endocrine, system of the body.

The endocrine system exercises control over a great range of body functions through the secretion of hormones. Hormones regulate everyday processes such as the turnover of calcium and the maintenance of blood sugar at the right level, and they also regulate long-term processes such as growth and reproduction. There is a close interplay between the brain and the endocrine system, since there are glands in the brain, such as the pituitary and the hypothalamus, which are stimulated by nerve cells and whose hormones act to regulate the secretions of other glands. For this reason, the complete hierarchy of hormonal control is known as the neuroendocrine system.

Because the neuroendocrine system plays a central role in keeping the body working smoothly, it has often been suggested that hormones hold the key to understanding how ageing is regulated. Some even believe that somewhere in the brain, perhaps in the hypothalamus, sits a clock that measures out our days. What this idea does not explain, however, is why such a clock should exist, and we have seen earlier that this is a very hard question to answer.

Undoubtedly, hormone levels do change with ageing, and it is reasonable to believe not only that the glands themselves age, but that changes in hormone secretions can have profound effects on the tissues that normally respond to these hormones, which may themselves also be ageing in ways that alter their responsiveness. Gross interference in hormone levels in rats – for example, by surgical removal of selected glands – produces extensive effects that can influence, among other things, how the animal ages. This is hardly surprising, but does it help us to understand how hormones interact with ageing in the unoperated animal?

We reminded ourselves at the beginning of this chapter that it does not make evolutionary sense to suppose that the ageing of our bodies is programmed. But this does not mean that our ageing process is entirely haphazard. We have already seen how the heart compensates for an increased workload by increasing the size of its left ventricle. There are many other examples of compensatory adjustments that occur in old age, the basis of which is to be found in the phenomenon of homeostasis.

Homeostasis is a term which describes the self-correcting mechanisms that keep things on an even keel. When the body gets too hot, we sweat and cool ourselves down. When we are too cold, we shiver and the muscle action warms us up. These are just two facets of the mechanisms that underlie temperature homeostasis. Life is unpredictable, but our bodies work best within a narrow envelope of physiological conditions. To keep us within this envelope, we have an impressive array of homeostatic mechanisms.

Natural selection did not design us to age, but it did design us to cope with all kinds of vicissitudes, such as illness, cold and hunger. Many of the challenges that occur as a result of ageing doubtless mimic other challenges that occur in younger individuals as a result of the sheer unpredictability of life. In time, the cumulative effects of ageing test even the cleverest of homeostatic mechanisms to the limit. This is why, as we age, we find that our adaptive response to physiological stress declines. Partly it is that the homeostatic mechanisms themselves do not work quite so well as they did when we were young; partly it is that we are using up some of their capacity already.

The ageing body is sophisticated, but it is not sophisticated enough. Throughout this chapter we have seen how damage at the cell and molecular levels seems to underlie the ageing of tissues and organs. Furthermore, we can see how some of the same processes, like oxidative damage by free radicals and the accumulation of junk proteins, crop up time and time again. Sometimes the result is what we call a disease, sometimes it is 'normal' ageing.

So, are we orchestras of organs playing to a score? There is no scripted finale that we can detect, and there is no conductor. But there are patterns in the rhythm of our ageing. Like the good jazz bands we are, we put up a pretty impressive performance. As we learn to read these patterns more clearly, we will be better able to tackle the problems of old age.

One pattern that emerges from all that we have seen so far is that our cells function less well as we grow older. But it is often said that it is the exception which proves the rule. The exception in this case is the disease that killed the vicar's wife: cancer. Cancers kill because malignant cells grow all too well. They are not frail but vigorous, yet they arise most commonly from older tissues. In the next chapter, we examine how we can fit cancer into our understanding of ageing.

CHAPTER 10

The cancer connection

Roses have thorns, and silver fountains mud,
Clouds and eclipses stain both moon and sun,
And loathsome canker lives in sweetest bud.

William Shakespeare, 'Sonnet No. 35'

We saw in Chapter 2 that some cancer specialists have questioned the link between cancer and ageing, but this is a minority view. There are very good reasons to regard the connection as a real one, and if we can but track it down, I believe we will have a better view of both of these important processes.

The first reason to link cancer with ageing is that many common cancers, including cancer of the lung, colon, breast and prostate, become very much more common in old age. Cancers have grown in importance as killing diseases precisely because we now live so much longer. Some cancers are extremely common. Prostate cancer is found in as many as one in two men over age 65, although this type of cancer can be relatively benign. Such cancers often develop slowly and remain undiagnosed, death occurring from some other cause in the meantime.

Similar patterns of cancer incidence are seen in developed and developing countries, but there are some differences. In regions like Navrongo, with shorter life expectancies, fewer people get cancers because other things may cause death first. There are also some cancers that appear to be linked to environmental factors that may vary from one part of the world to another. Infection with the

Epstein-Barr virus that causes glandular fever has no known association with cancer in the developed world, but in Africa it is linked with a cancer of the lymphatic system and in China and South-East Asia it is linked with cancer of the back of the nasal passages. Why these links between the virus and these two kinds of cancer should exist is unknown, but it suggests an interaction with something else in the environment.

The overall death rate from all kinds of cancer shows a steady increase until age 90. After age 90 the rate levels off, and at older ages it even declines. The reason is probably just that tumours, which are aggressive, energy-hungry growths, cannot suborn the tissues of a very old person to serve their needs as easily as they can the more vigorous tissues of a younger person.

Not all cancers, of course, are linked to old age. Incidence of testicular cancer peaks in the thirties and declines sharply thereafter. Certain rare cancers like retinoblastoma, which affects the eye, are found almost exclusively in children. Leukaemias – cancers of the blood-forming cells – can strike at any age but certain types are commonest in children and young adults.

A second reason to link cancer and ageing is that both are thought to arise from cellular damage. Mutations to DNA are known to play a part in cancer formation and, as we saw in Chapter 8, mutations are probably a contributing factor in ageing. It is an interesting fact that mice and humans have about the same lifetime risk of getting cancer, in spite of the fact that the human life span is 30–40 times as long as that of a mouse. When you factor in that a human has many more cells than a mouse, any of which could become malignant, it is clear that the risk of starting a cancer per cell per day is much, much higher for mouse cells. The difference is explained by the observation that human cells repair damage, including damage to their DNA, a lot better than mouse cells. Disposable we may be, but we are a lot less disposable than a mouse.

The third reason to link cancer and ageing is the most tantalising of all. We saw in Chapter 7 that cells grown from normal human tissue have finite replicative life spans in culture. Cells from cancers grow without limit.

In 1952, a woman who had the misfortune to suffer from a cancer of the cervix was treated at Johns Hopkins medical school in Baltimore, Maryland. The woman's name was Henrietta Lacks. A small piece of her tumour was used to start a culture of cells, using the cell culture procedure described in Chapter 7. The biopsy was placed in a culture dish, cells were observed to grow out from it, and these cells were subcultivated repeatedly.

The cell line grew well. In fact, the cells did not stop growing. The cell line was given a name, HeLa, taken from the first two letters of Henrietta Lacks' first and last names. Remember that in the 1950s the Hayflick Limit had not been discovered – this would not occur until 1961 – so no one was surprised that the HeLa cells did not stop growing. However, HeLa was, as far as we know, the first 'immortal' human cell line.

HeLa grows so well that, over the last half-century, samples have been passed from laboratory to laboratory around the world, where they have been used in countless scientific experiments. They are still used today. Nearly any tissue culture laboratory of any size is likely to have a few HeLa cells stored in its freezer. Many more HeLa cells have been grown than ever existed in Henrietta Lacks.

To my mind there is a deep poignancy in the fact that Henrietta Lacks' cancer killed her long ago, but her cells live on, and on, and on. This is far from the non-dying kind of immortality that Woody Allen says he hankers after. Nor is it the kind of immortality enjoyed by the likes of Einstein and Pasteur, whose names are for ever linked to their discoveries, as perhaps Hayflick's will be after ageing exacts its toll. Maybe it is rather fun as a gerontologist to be known for a 'Limit'. But HeLa has definitely conferred on Henrietta Lacks immortality of a kind.

Just what does cellular immortality mean? And how is it linked to the Hayflick Limit that we see in normal cells? To answer these questions we need to take a closer look at cancers.

One of the scariest things about cancer is that it takes only one rogue cell, out of all the myriad cells in our bodies, to start up a growth that may kill us. Recall that there are around 100,000,000,000,000 cells in the average human adult. This is alarming, but when you think about it, the really remarkable thing is that our cells do not become cancerous more often. A cell, after all, is a little life form of its own, and life forms are subject to the law of natural selection. However much we may suffer if one of our cells breaks free from the normal controls on its growth and replicates without restraint, the cell is actually doing nothing more than following the imperative of natural selection acting at the cellular level.

Our cells, as we saw in earlier chapters, are biochemical factories containing their own genetic blueprints and their own molecular machinery for working from these blueprints to make the components that they need in order to function. Inside the body, cells are supplied with nutrients and raw materials from the circulating blood. They dump their waste products into the bloodstream to be carried away and excreted. As long as this cosy environment persists, cells do their own thing, responding as necessary to hormones and other signals that reach them from near or distant cellular neighbours. But if a cell is capable of division, and if that cell starts to divide more than it should, then natural selection at the cell level will, in the short term, favour its continued replication. A mutation that gives the cell a reproductive advantage over its fellows will be copied into more daughter cells than the unmutated gene in the normal cell neighbours. Natural selection, the friend that made us, becomes natural selection, the foe that may destroy us. Cancer cells fight hard against our efforts to destroy them with chemotherapy and radiotherapy. If we are unlucky, they adapt – that is, evolve – and escape.

In the early 1980s a novel class of genes started to make headlines in the scientific and lay press. The genes were called oncogenes, or genes for cancer. The first oncogenes were identified as genes carried by certain oncogenic, or cancer-causing, viruses.

When an oncogenic virus enters a host cell, the oncogene seems somehow able to make the cell cancerous.

In a short while, it was discovered that most viral oncogenes were not novel genes, peculiar to the virus, but mutated copies of normal cellular genes, which had been picked up at some point in the evolution of the virus gene sequence, probably as an accident of genetic recombination during the virus's sojourn inside a host cell. Some of the oncogenes turned out to be genes that code for cellular growth factors.

A growth factor is a cellular signal that tells a cell to divide. Mostly these signals come from other cells. This provides a safe system of checks and balances, like those that prudent investment banks use to safeguard the flow of funds. But if a dividing cell starts to produce and respond to its own supply of growth factor, and if its daughter cells inherit this trick, then the dangerous result is a clone of cells that expands much too fast. When Nick Leeson brought about the downfall of Barings Bank in 1995, it was because he had too much autonomy for his actions.

After the initial discovery of viral oncogenes, a rush of other oncogenes were identified. Nowadays there are dozens of genes that carry the oncogene label. Many of these are perfectly normal genes that regulate aspects of cell division and cell death, but which when they do not work properly, or if they are on when they should be off, have been implicated in the development of cancer.

As well as oncogenes, another category of genes has been found to exist, called tumour suppressor genes. These are genes that act to prevent cancer – for example, by blocking cell division or by causing the suicide of a damaged cell. The twin concepts of tumour suppressor genes and oncogenes describe the yin and yang, or stop–go nature, of the control of cell proliferation.

It is easy enough to see that breaking out from normal controls on cell growth and division can lead to excess cell proliferation. But what happens next? And where does cellular immortalisation fit in?

Early on in the scientific study of cancer, it was realised that to go from a normal cell to a full-blown malignancy requires more than a single step. A cell that produces its own growth factor can give rise to an expanding clone of daughter cells, but most cancers are more than just a clonal overgrowth. An expanding tumour must develop its own blood supply and invade surrounding tissue, otherwise it will spontaneously regress and die. The most dangerous of cancers, such as malignant melanomas, acquire the ability to metastasise – that is, to release daughter cells that seed new tumours at distant sites within the body.

From observations on the step-wise nature of tumour progression, and calculations based on epidemiological data concerning the rates at which new cancers arise, came the idea of cancer as a 'multihit' process. Several changes must accumulate within a single cell before a malignancy occurs. Most of these changes are thought to be somatic mutations – that is, alterations to the DNA sequence that occur in a single somatic cell, as opposed to germ-line mutations that arise in a germ cell and affect the next generation. Somatic mutations are what Leo Szilard suggested might cause ageing.

Mutations are known to cause cancer. There can be inherited mutations that confer increased risk of certain kinds of cancer, as we shall see very soon. But the kinds of somatic mutation that Szilard had in mind are not inherited; they happen in the DNA of individual cells within the body. It only requires the right (or wrong!) mutations to occur in a single cell to cause a malignancy.

Certain chemicals and ionising radiations that cause mutations result in general increases in cancer incidence in experimental animals. Survivors of the atomic bombings at Hiroshima and Nagasaki developed cancers more than the general population. A widely used test to screen for potential carcinogens is the Ames test, developed by the scientist Bruce Ames at the University of California, Berkeley. In fact, what the Ames test measures are mutations rather than cancers, but it has proved to be an invaluable surrogate.

When we put the idea of oncogenes and tumour suppressor genes together with the observation that mutations cause cancers, it is easy to explain the multihit nature of cancer. All we need to do is suppose that a cell needs to accumulate mutations in one or more oncogenes that promote cell growth, and acquire mutations in one or more tumour suppressor genes that inactivate anticancer defences. Then the cell becomes malignant. Most human cancers are found to carry mutations in oncogenes and tumour suppressor genes. The molecular analysis of just which of these genes are mutated can help cancer specialists predict the future course of the disease.

We can see, incidentally, how the accumulation of mutations is facilitated if some of the early hits cause a cell to increase its division rate. If the gene mutation rate is low – say, one in a million per month – then the time taken to acquire ordinary mutations in two given genes in the *same* cell will be long. But if the first mutation causes the mutant cell to divide more than its neighbours, then because its daughter cells all carry copies of the first mutation, the fraction of cells at risk of picking up the second mutation is that much greater.

A particularly clear example of how mutation in a tumour suppressor gene can be linked to cancer is found in the case of retinoblastoma, mentioned earlier as a rare cancer of the eye affecting young children. Retinoblastoma occurs in two ways, one hereditary, the other not. All of us normally carry two copies of the *RB1* gene, a tumour suppressor gene, one copy coming from each of our parents. A single intact copy of the gene is enough for the tumour-suppressing action to work, but if both copies are mutated, a cancer becomes likely.

In non-hereditary retinoblastoma, the child at conception receives intact copies of the *RB1* gene from both its parents, but through bad luck, one of the cells in its retinas acquires mutations in both copies and a cancer forms. Because the risk that this happens is tiny, the likelihood of both eyes being affected in this way is negligible. Non-hereditary retinoblastoma almost always affects only one eye.

In hereditary retinoblastoma, however, a child at conception receives one copy of the *RB1* gene that is already mutated. This makes it much more likely that some retinal cells in the developing eyes will, by chance, suffer mutation in the remaining intact copy of the gene. In cases of hereditary retinoblastoma, it is common for both eyes to be affected.

Let us take stock of the story so far. We have seen how the cancer rate rises with age. We have seen how cells grown from malignant tumours grow indefinitely. We have seen how oncogenes and tumour suppressor genes come into the story, and we have seen how mutations in these genes can contribute to the multistep progression to cancer. We will now look a little more closely at the difference between immortal cancer cells and mortal normal cells, and we will begin by looking at a strange kind of growth called a teratoma.

Normally, as everyone knows, an egg cell needs to be fertilised by a sperm before it begins to develop into an embryo. But every once in a while, an egg cell can become activated while it is still in the ovary and start to develop all on its own.[14] The result, in mammals, is not a virgin birth but a teratoma. The egg divides and begins the early stages of embryogenesis apparently normally, but something is lacking and it fails to complete the proper developmental sequence. What happens instead is that the embryo becomes disorganised and forms a shapeless mass of cells which, on close inspection, contains a bewildering variety of different cell types and even partly formed organs. A teratoma may contain bones, skin, bits of glands and even hair. It is also possible for teratomas to form in men as well as women, from the sperm-forming cells of the testis.

As if teratomas were not ghastly enough, they can also develop into a life-threatening form of cancer called a teratocarcinoma. Teratocarcinomas grow without limit until they kill the host. In laboratory rodents, it has been shown that a teratocarcinoma can be transplanted from animal to animal of the same genetic strain, and in each new recipient the cancer will grow to kill its host. In this respect, a teratocarcinoma is just like any other aggressive cancer.

But teratocarcinomas have one astonishing difference. If some cells are taken from the teratocarcinoma of a mouse, and if these cancerous cells are then injected into an early-stage mouse embryo, the resulting animal is entirely normal! You might think that the cancer cells are simply eliminated from the developing embryo, but this is not the case: if they are suitably marked, they can be found contributing to the tissues of the adult mouse. The only explanation is that the teratocarcinoma cells have been tamed by the powerful developmental signals that are being produced in the early-stage embryo.

In fact, teratocarcinoma cells seem to be normal cells that have lost their way, like a member of a chorus line who muddles his cues and wreaks havoc on the dance routine, only to deadlier effect. But even though teratocarcinoma cells appear normal, they are far from being typical cells. In culture, the population as a whole may grow indefinitely, but individual cells can be found to turn into any of a huge variety of specialised cell types, reminiscent of the *mélange* of tissues seen in a teratoma, and to lose their immortality. To maintain a pure culture of teratocarcinoma cells requires special culture medium. And this is not all that is strange about these cells. If you take a perfectly normal, early-stage mouse embryo, fertilised in the usual way, dissociate its cells and grow them in culture, you get a cell culture that is almost indistinguishable from teratocarcinoma cells.

What teratocarcinomas tell us about cancers are two very important messages: you do not have to have mutations to get a cancer; and cancer cells can behave very much like the cells of an early embryo. In fact, the second of these points was recognised a long time ago when the word 'neoplasia' was coined to describe tumour growth. Neoplasia refers to a property that many cancer cells show: namely, that they revert to a more juvenile state of cellular development. They are nearer to the germ-line.

When I first realised why ageing occurs, back in my bath in 1977,

I was puzzled by the immortality of cancer cells and its similarity to the immortality of the germ-line. Was it possible, I suggested in my paper in *Nature*, that the conversion of normal somatic cells to a cancerous state involved the accidental reactivation of mechanisms for germ-line immortality that were meant to be switched off in the disposable soma?

In the late 1980s, an interesting discovery was made by a Canadian cell biologist, Calvin Harley, now research director at Geron Corporation, a biotechnology company with particular interests in ageing and cancer. Harley found that, when human cells are grown in culture, their telomeres get shorter and shorter. Telomeres are special DNA structures at the ends of chromosomes which are thought to prevent the chromosomes from unravelling or from accidentally joining up with each other. Telomeres have been likened to the plastic tips at the ends of shoelaces, without which the shoelace frays rather quickly.

But if telomeres always got shorter when cells divided, as in the cells Harley grew, the germ-line would soon be in trouble. After a number of generations, the telomeres would disappear completely. This disaster is prevented by a special-purpose enzyme called telomerase, which is found in germ cells and which fixes the ends of the chromosomes in an unusual way. When the DNA in the chromosome gets copied, the ends of chromosomes present a problem. The molecular machinery that replicates the DNA attaches to the strand of the helix it is copying and adds new bases of DNA (according to the pairing rules described in Chapter 8) as it moves along. But when it gets to the end of the strand, it cannot copy the very last section for the simple reason that it cannot, at the same time, hang on to the terminal sequence *and* copy it. Telomerase makes it possible to copy the strand right to its end because, in effect, the enzyme has an extension piece that allows DNA replication to continue to completion.

The DNA 'end-replication problem' was recognised as early as 1971 by a Russian scientist, Alexei Olovnikov, who with great insight suggested that it might have something to do with ageing.

But it was not until the last decade that the telomere story has really become exciting. Without the telomerase enzyme being present, the cell's inevitable failure to replicate the DNA in its chromosomes right to the end means that, with each cell division, the telomeres get a little bit shorter. When the telomeres finally get too short, the cell is in trouble. Exactly what goes wrong we do not know, but it may be that genes near the ends of the chromosomes – near where the telomeres begin – no longer work properly. Or it may be that the chromosomes stick together in inappropriate ways that interfere with cell division.

Telomerase is off in somatic cells and on in germ cells, which is how the telomeres of the germ-line are protected from growing ever shorter. The extraordinary thing is that, if we look at cancer cells, we find, nearly always, that telomerase is turned back on. In other words, we have a perfect example of a germ cell maintenance system that gets shut off in the soma but accidentally reactivated in cancer, just as I predicted in 1977.

The excitement of the telomere discoveries is their potential relevance to cancer *and* ageing. However, we should not exaggerate the significance of telomeres and telomerase because there are some snags. For a start, telomeres get shorter only in dividing cells and therefore do not shorten in brain and muscle. Yet these organs certainly age. Furthermore, mice have much longer telomeres than humans, which is not at all what you would expect if the telomeres are like fuses that burn down to detonate an ageing 'time-bomb'. In 1997, a report was published describing a genetically engineered mouse strain that has no gene to make telomerase. If telomerase activation were the only thing involved in cancer, these mice should not have developed tumours. But they did, and at the same rate as normal mice. On the positive side, a study with cultured cells showed in 1998 that introducing telomerase into normal cells (in which telomerase is switched off) enabled them to grow through the Hayflick Limit. This indicates that, even if telomere shortening and telomerase are not the keys to unlocking all of the secrets of ageing and cancer, as many hoped, the research may have considerable practical value if it allows normal cells to be grown more easily for use in drug development and possible future cell replacement therapies.

Teratomas and telomeres are pieces in the bigger jigsaw puzzle of just how cancer and ageing fit together. Mutations cause cancer, but you do not *always* need mutations for a tumour to form, so what is it that you do need? It seems that you need something which abrogates normal cell senescence and causes cells to revert to a state that is something like the germ-line. This implies that the switching on and off of genes is a part of the process. After all, each cell in your body contains the same genes. The reason why one cell may be different from another is that their gene switches are set differently as the organism develops from the fertilised egg. Mutations can break switches, turning genes on or off in inappropriate contexts. In the case of teratocarcinoma cells, however, it seems that the switch mechanisms just get confused without necessarily getting broken. The latter can happen by epimutations – that is, the disorganisation of DNA tags, or Post-It notes, that we encountered in Chapter 8, and which can reversibly turn genes on and off. But what are the genes whose switching is so important in cancer?

We saw in Chapter 7 how one school of thought holds that cell senescence is an anti-cancer mechanism. At first sight, this is a seductive idea. In order to protect the organism against cancer, or so the theory goes, genes evolved to limit cell proliferation. This is the good news. The bad news is that cells age, and that eventually this contributes to the ageing of the organism. In other words, the genes for cell ageing are part of an evolutionary trade-off, in which protection against cancer is the benefit and ageing the cost.

I do not accept this idea completely. It is too teleological, or purposive. For a start, the Hayflick Limit in human cells is long. Remember that fifty or even just thirty cell doublings produce a huge number of cells. The idea that tumours keep starting out and then bumping into the Limit after they have gone through this many cell divisions seems to me just a little implausible. If it happened often, the abortive tumours would be rather obvious, yet we do not see them. It also seems a risky strategy to let the potentially malignant cell go through so many rounds of cell division before slamming the brakes on. The greater the number of cell divisions, the greater the chance of acquiring mutations that will break the hypothetical machinery that stops the cells in their tracks. We know cell division is dangerous in this way from epidemiological studies on oral contraceptives and their link with breast cancer. Epidemiologists think that it is precisely because oestrogen stimulates the cells in the breast to grow that it increases the risk of accumulating the cancer-causing changes.

The best kind of anti-cancer defence is one that stops rogue cells dead in their tracks just as soon as possible, which is what some tumour-suppressor genes do when they cause damaged cells to commit suicide, or what the immune system does if it detects that a cell has gone wrong and destroys it. I read recently the remarkable story of a dog that saved its owner from a malignant melanoma because it sensed that the growth smelled odd and worried at it, prompting her to seek a medical opinion. Early detection of trouble, by whatever means, is the way to go.

All right, you might say, but what if the front-line defences fail? Maybe you need the Hayflick Limit as a fail-safe. Well, perhaps. But it is a bit curious, isn't it, that we humans have such a distant fail-safe mechanism (50–60 cell doublings), when mice have a much shorter Hayflick Limit (10–15 cell doublings)? On the face of it, a shorter Limit gives tighter control, which is what we with our long life spans need a lot more than mice. But actually it doesn't, because mouse cells turn cancerous much more easily.

We find it so hard to let go of the idea that ageing has a purpose that perhaps we are too easily persuaded by the notion that ageing is an anti-cancer mechanism. It is not. And yet there is a real connection between ageing and cancer, which has, I believe, much to do with the fundamental distinction between the germ-line and the soma. Somatic cells are cheaply made and disposable, but each somatic cell contains within itself the genetic wherewithal to become germ-like again. Cancer is an accidental reversion to a germ-like state. The same general mechanisms that protect against cancer protect against ageing. This is why long-lived species, with their better cellular protection, get cancer later than short-lived species.

CHAPTER 11

Menopause and the big bang

The enormous proliferation of menopause literature belies the utter lack of understanding of what
is really going on.

Germaine Greer, *The Change*

In the continent of Australia there are some remarkable creatures,
none more so than the duck-billed platypus, a mammal that lays
eggs, has a beak, and swims under water with the ease and grace of
a fish. When European colonists first encountered the platypus and
sent a specimen home, it was taken by naturalists for a crude hoax.
As well as the platypus and other delights, like the kangaroo, or
horrors, like the brown snake and funnel-web spider, there is an
unobtrusive little marsupial mouse which has a rather unusual
habit that it managed to keep hidden from scientists until the
1960s. Males of the species *Antechinus stuartii* indulge in 'big
bang' reproduction.

Males in many species show a considerable preoccupation with
sex, but male marsupial mice take matters to an extreme. As the
August mating season draws near, testosterone levels build steadily
higher, reaching lift-off in late July. At the same time, the adrenal
glands greatly enlarge, sending elevated levels of corticosteroid
hormones flooding into the blood stream. These are signs that the
males are entering a state of extreme physiological excitement and
stress. They soon launch into violent battles with one another for

the opportunity to mate with the females (who are also getting rather worked up, but on the whole are taking things more calmly).

By the time the fun and games are over, the males are in a sorry state. In addition to the scars of battle, many of them have serious stomach ulcers that bleed severely. Their immune systems are shot to pieces with the result that they fall easy prey to parasites. Nearly all of them will die in the course of the next few days, worn out by the ravages of sexual frenzy. The females survive to raise and suckle their fatherless young, but many of them will end up quite frail as well. Only a few of the females survive to breed again the next year.

Big bang reproduction – more properly known as semelparous reproduction, which is basically Latin for having all your babies at once – is not unique to males, or to *Antechinus*. Male and female Pacific salmon, after a long and arduous journey across ocean and up river to their spawning grounds in the lakes of the northern United States and Canada, breed just once and die. The lakes fill with the corpses of parents, but life is renewed and in time the salmon hatchlings will emerge from the eggs and swim away downstream, carrying with them a chemical memory of their birthplace, so that they too can return one day to give new life and die. Beneath the waters of the Mediterranean Sea, the female octopus lays her eggs in the hole in the rock that serves as home. She then fans water across the egg cluster to keep them oxygenated and loses interest in feeding herself. The eggs hatch and the mother octopus dies soon afterwards, never regaining the drive to live.

It is possible to weave a strand of heroic pathos into the stories of the salmon and octopus, so let us counter this with the grisly tale of a little mite of the genus *Adactylidium*, whose young hatch inside the body of the mother and eat their way out, killing Mum in the process.

The thing that makes semelparous reproduction so intriguing is that, in most cases, there is no obvious need for the sacrifice of the parent. The case of the mother mite is an exception because, although serving oneself up as breakfast goes beyond the normal demands of parenting, no one can deny that it gives the little mites a good start in life to be sent into the world with a full tummy. In the case of the mother octopus, it is not at all clear why she does not resume normal feeding when the little octopuses hatch. It is not as though she is rushed off her feet – all eight of them – ministering to her little ones' needs. All she does is die.

In fact, the mother octopus *can* go on living, but a drastic measure is needed. It has been found that what makes the mother lose interest in food is a hormone secreted from her optic gland. Surgical removal of the optic gland leads to cessation of broodiness, resumption of feeding and greatly extended survival. In marsupial mice it is also a gland, in this case the adrenal gland and the hormones it produces, that causes the death of males after mating. If death is preventable, why do these hormones not get switched off when mating is over and before the animal dies?

We know of another case where hormones do something surprising, and this is the menopause, which occurs in our own species in women around age 50. In spite of claims to the contrary, there is no such thing as a male menopause. Men lose fertility due to age changes in their testes and they become increasingly prone to impotence – failure to produce and maintain an erection – but they do not show a specific shut-down of reproductive function.

Menopause is the final menstrual cycle and it signals that reproduction has ended. Yet a woman at age 50 is often in excellent physical condition and may live as long after her menopause as all her reproductive years before it. In the last few years, we have begun to see cases where, through the use of medically assisted fertility techniques, women past the age of menopause have given birth to healthy children with no major complications. All of this makes the menopause a puzzle. In Darwinian terms, it seems an odd thing to shut down reproductive function so early.

The menopause and big bang reproduction are instances of biological behaviours that seem to run counter to the principle that natural selection acts always to maximise reproductive potential. As such they are often linked, and both may be held up as counter-examples to the arguments against the programme theory of ageing in Chapter 5.

In truth, menopause and big bang reproduction have very different origins. The rapid post-reproductive death of a semelparous animal occurs because for big bang reproducers, natural selection has little interest in what happens to the organism after it has completed its reproduction. Menopause, by contrast, is thought to occur because after a certain age, women become too valuable to their kin groups to risk in having further babies of their own. We will look more closely at the explanations of semelparous death and menopause in the remainder of this chapter.

To understand the apparent programming of death in a semelparous species, we need to consider why a species might evolve big bang reproduction in the first place. Like many choices in life, it boils down to balancing levels of risk and the expected return on an investment. The Hollywood film industry is a good example.

When I was young, new movies came to our screens rather gradually. After opening, usually in London, they made their way through a succession of major, intermediate and minor venues around the country, often stopping only a week or two at each cinema. If a movie was a hit, it might make the rounds again. Because the release was staggered in this way, any advertising was limited and local. Nowadays a big film opens nationwide on the same day, often with a huge advertising build-up to maximise the early financial return. After the first run, it is rare for a film to return to the cinemas again. Secondary sales are through television and video release.

One must suppose that the moguls of the film industry found it more cost-effective to operate their movie releases on a semelparous basis. Major effort and advertising investment is directed at maximising the box office takings from what is usually the film's one and only outing to the cinema.

In much the same way, genes have two broad options in terms of reproductive strategy. Either they stake all on a single shot at semelparous reproduction, the big bang strategy, or they go for distributing reproduction over repeated reproductive bouts, the so-called iteroparous strategy. Each strategy has its advantages and disadvantages. The major drawback of the iteroparous strategy is that, although you hedge your bets and lessen the risk of reproducing at the wrong time, you have to keep a lot in reserve in order to have a reasonable chance of surviving to breed again. As well as investing in reproduction, you need to invest in maintenance for the longer term. Semelparous breeders have no such inhibitions. They go for broke.

There is an old joke about a sexologist who carried out a study on the link between frequency of sexual intercourse and happiness. A roomful of couples were asked who had sex every day. A few couples raised their hands, looking smug and very happy. The sexologist then asked who had sex once a week. More couples raised their hands, looking quite happy. The sexologist next asked who had sex once a month. Further couples raised their hands, no longer looking very happy. Finally, the sexologist asked if there was anyone who had sex just once a year. A single couple raised their hands, beaming all over their faces. Puzzled, the sexologist asked why they were so happy. The answer (you guessed it): 'Today's the day!'

Semelparous reproduction is a bit like this. Most of the life span is spent getting ready for the big event. The animal grows, stores energy and prepares its gonads for one explosive bout of reproduction. When the signal is given, and such signals are nearly always hormonal, there is no holding back. Resources are mobilised to maximise reproductive effort, even if this leaves the animal so damaged and depleted that it dies soon after.

Because death in semelparous species follows hard on the heels of reproduction, which itself is hormonally driven, it can easily look as if death itself is programmed. Usually this is not the case. Death serves no particular purpose; it is either a side-effect of the overwhelming drive for reproduction, or simply the result of neglect. Just occasionally, as in the case of the mother mite, natural selection goes a stage further and puts the death of the parent to good use. Thus, we can see that death in semelparous species is not an exception to the ideas of the disposable soma theory, encountered in Chapter 6, but an extreme instance of it.

We will now look at how the evolution of the menopause is to be explained. Menopause, or 'the change', is something that is unique, or very nearly so, to human women. Female chimpanzees and macaques show something a little like it, but the shut-down of reproduction in chimpanzees and macaques occurs nearer to the end of the species' maximum life span, not around half way through it, as in women, and is less clearly defined. Female pilot whales are also thought to have a kind of menopause, but for obvious reasons, this process is a little tricky to observe.

The trigger for menopause is that a woman's ovaries start to run out of eggs. Unlike sperm, which are produced on a continual basis following puberty, all of the eggs a baby girl will ever have are produced before she is even born. The store is finite and egg numbers decline all the while, even from birth. At its peak, when the female foetus is 4-5 months old, the egg number can be as high as 7 million but only 1 million remain by the time the baby girl is born. By the time of puberty, the number has dropped to about quarter of a million. By her mid-thirties, just 25,000 eggs remain. From then on, the rate of egg loss accelerates, and by age 50 the last of the eggs are pretty much gone. For the modern woman, who might not complete her education much before 30, the window of time for child-bearing can be cruelly short.

A bit of simple arithmetic shows that egg loss cannot be accounted for solely by the rate at which eggs get used up in ovulation. Thirteen menstrual cycles a year times 36 years between puberty and menopause is just 468 eggs. And if a woman gets pregnant, or uses oral contraceptives, the total number of cycles will be even smaller. This leaves more than 99.9 per cent of eggs unaccounted for.

It is worth speculating for a moment about what all that extra egg loss signifies. It may be that what we witness when we see more than a thousand eggs being destroyed for every one that is released from a mature follicle, and thus given an opportunity to enter the reproductive cycle, is a stringent form of quality control designed to keep the female germ-line from ageing. Eggs-in-waiting in the ovary are surrounded by special nurse cells within the follicle. It may be that the egg or its nurse cells are on the lookout for damage, and that the egg is destroyed if damage is detected. This might explain why the rate of egg loss accelerates towards the end, if the kinds of molecular damage we considered in Chapter 8 are beginning to build up in eggs that by now may be 40–50 years old. And it might also explain why the frequency of Down's syndrome and other genetic abnormalities increases with maternal age, if the mechanisms for detecting and eliminating damaged eggs begin to work less well.

While the eggs last, the regular monthly cycle is co-ordinated by a dialogue between hormones secreted from the pituitary gland in the brain, and hormones secreted by the ovary. A pituitary hormone called follicle-stimulating hormone results in the activation of a chosen follicle, and a subsequent burst of another pituitary hormone called luteinising hormone causes the follicle to burst and release the egg into the fallopian tube. Meanwhile, the nurse cells of the activated follicle produce the ovarian hormones progesterone and oestrogen, which in a normal cycle prepare the wall of the uterus to receive a fertilised egg. If the egg remains unfertilised or fails to implant, the blood-engorged wall of the uterus breaks down and is shed as menstrual flow. And the cycle begins again.

With ageing, the tight co-ordination between the pituitary and the ovary begins to loosen and the cycling becomes less regular. A similar process affects all female mammals, whether they be shrews, sheep or she-elephants. In all species, the store of eggs declines and may eventually be exhausted, but the store is generally large enough to last lifelong in the wild. Only in humans do the eggs run out so long before the end of the species' life span.[15] When the eggs finally run out, a woman enters the 'change', or menopause. The nurse cells in her ovaries find themselves to be redundant, and they stop producing progesterone and oestrogen completely. A little bit of oestrogen is still produced from elsewhere in the body, but the level overall drops to something less than a twentieth of the premenopausal level. The pituitary, incidentally, increases production of follicle-stimulating hormone and luteinising hormone in an attempt to kick the ovaries into cycle again, which is the glandular equivalent of yelling over the telephone at someone who is deaf. So much for the 'wisdom' of the ageing body.

As is well known, the menopause is accompanied by a number of discomforting and unpleasant symptoms, such as hot flushes, sweating, insomnia, dizziness and palpitations. Many of these are due to small perturbations in the finely tuned mechanisms for temperature control, which seem to be provoked by the marked reduction in the levels of oestrogen. Most of these symptoms are temporary, but longer-lasting damage to bones and arteries is beginning to happen unnoticed, which may not become apparent for many years.

So far, there does not seem to be much to be said in favour of the menopause. But you might feel differently if you had been a Victorian mother with sixteen children wondering if at age 50 you could face going through the business of pregnancy yet again. You might also feel differently if you were a mother in present-day Navrongo wondering the same thing. Not least, you might worry about what will happen to the youngest children that you have already, or to your daughter or daughter-in-law who is pregnant for the first time and needs your assistance, if you should die in childbirth.

In much of West Africa, more women die in childbirth than from all other kinds of accident. The lifetime risk of dying of a pregnancy-related cause in the developing world is between 1 in 15 and 1 in 50, compared with an average lifetime risk of between 1 in 4,000 and 1 in 10,000 for a woman in the developed world. Small wonder that the World Health Organisation has described the way that maternal mortality continues to reap its horrifying harvest as 'the shame of the century'.

And remember, these figures for maternal mortality are what we see *after* menopause has set a cut-off in the age of pregnancy at age 50, with reduced fertility for some years beforehand. Maternal mortality rises with age, so without menopause the death toll would be even greater. In fact, menopause regularly occurs 5 or more years earlier in poorer parts of the world, where foetal malnourishment results in reduced numbers of eggs being formed, so the protection offered by menopause is greater still in these most challenging of circumstances.

Against this background, it seems reasonable to suppose that menopause is not entirely a bad thing and that perhaps it evolved to fulfil a positive role. Because menopause is peculiar to our species, we need to ask ourselves what is special about human biology that might have favoured its evolution.

Well, the first thing to note is that we live unusually long lives. Humans have the longest life spans of any mammal. It could be, as some have suggested, that modern life simply allows women to outlive an egg supply that would have lasted lifelong during most of human evolution. Or it could be that the human egg store has not kept pace with our evolving longevity because eggs cannot keep for ever – that is, their 'shelf-life' is limited.

There is a simple counter to the limited shelf-life suggestion if we consider the she-elephant. A she-elephant remains fertile at least to age 55, which is pretty close to her maximum life span. Her eggs are formed before birth too. How come they can keep for nearly 60 years if ours cannot?

To understand menopause, we need to look a bit more closely at why it is that we humans live so long. The driving force behind human longevity has doubtless been the evolution of a big brain. As the brain got bigger, our ancestors got smarter, and as they got smarter, they were better able to control their level of environmental risk. This will have put selection pressure on their somatic cell maintenance systems to work at higher levels, as we saw in Chapter 6. There is no point in evolving a brain that helps you survive the hazards of your environment, if you do not also evolve a more durable soma.

As more resources were directed into somatic maintenance, fewer were available for reproduction. But this was all right because one of the things about having a bigger brain is that it takes longer to grow and programme it. Brains became larger, life span got longer, somatic repair got better, and reproduction came to be a little slower. And into the bargain, our ancestral species found that living in social groups made good sense because there was safety in numbers, especially if those around you were your kin, who would have a genetic interest in your survival.[16]

Now comes the interesting part, because the general package grew to be so good that the brain got bigger, and bigger, and bigger! A big head on a baby is a bit of a problem when it comes to giving birth, as any mother knows to her discomfort. The problem was compounded by the fact that over time our ancestors stopped running around on all fours and began walking on two legs. The mechanics of walking upright imposed a constraint on pelvis size which meant that the channel through which the baby was born could not get any bigger. The solution that evolution found to this tricky problem was that human babies, unlike other mammals, came to be born with their brains half-grown.

It takes the best part of a year after being born before a baby's brain finishes its growth, during which the infant is wholly dependent on its kin, usually its mother, for survival. Its takes a further period of several years for the child to learn the language and social skills of its group. During this time, if its mother dies, the child is much less likely to survive.

Add to this the fact that an older mother, by around age 50, is already beginning to experience some of the adverse effects of ageing and suddenly it begins to make sense to suppose that she will actually enhance her genetic contribution to future generations if she stops having more babies of her own and thereby increases the likelihood that she will survive to raise her later-born children and to assist with her own grandchildren.

One might also add that by the time a woman has reached age 50 she will have acquired a considerable wealth of experience that will enhance her value to her kin group. It is perhaps no surprise that in traditional West African societies, such as exist in and around Navrongo, post-menopausal women cast off the restrictions associated with women of reproductive age and attain a new status as honoured members of the community.

A rather crucial question, upon which the plausibility of this evolutionary explanation of menopause depends, is how often our ancestors actually lived long enough to experience menopause. If hardly anyone lived past age 50, the idea that menopause evolved for the reasons suggested above seems unlikely. But we must remember that, although life expectancy at birth in times past was short, those who made it through to their thirties and forties could often expect to live another 30 years. The few records that are available from medieval times suggest that as many as a quarter of the population reached menopausal age. Major changes in population structure did not begin until the seventeenth and eighteenth centuries, so this pattern may have existed in earlier periods as well and would have been quite enough to provide the basis for natural selection to act upon.

We can also get some idea of the likelihood of survival to age of menopause under conditions somewhat similar to our evolutionary past from studies of present-day hunter gatherer communities. One particularly illuminating study is of the Ache people, a small indigenous population of hunters and gatherers who live in the rain-forest of eastern Paraguay. The Ache were visited and observed between 1978 and 1995 by anthropologists Kim Hill and Angelina Hurtado from the University of New Mexico and their colleagues. From their extensive data it is clear that, in spite of high infant mortality, six out of ten Ache females survive to the average age of first birth, which is between 19 and 20, and they then have a further life expectancy of about 40 years. In other words, many of them will live through menopause.

The studies of the Ache also support the idea that children whose mothers die face a bleak future. Hill and Hurtado write moving descriptions of the fate of Ache orphans, who are often killed by the group. In one particularly poignant interview, an Ache man described how he killed a 13-year-old girl whose mother had died in an epidemic. In Hill and Hurtado's words:

> The man told the story with tears welling up in his eyes and explained that it was the Ache custom to kill children after their parents died. We were distressed by the interview and couldn't help berating the man for what seemed like inhuman behaviour (we had heard many tales of child homicide and had even been present in Ache camps when some small children were suffocated but we had never heard such gruesome details of the sacrifice of an older child who was described as beautiful, healthy, and happy). The killer asked for our forgiveness and acknowledged that he should never have carried out the task and simply 'wasn't thinking'. He finally explained that 'the old powerful men told us we had to kill all the orphans, and we did as they said without thinking'.

Infanticide of orphans, which may well have occurred in other peoples, can only have strengthened the force of natural selection for menopause, in order to minimise the risk of leaving one's children in this perilous state.

The same kind of worry is seen today when post-menopausal women are assisted in bearing children. Is it fair, many of us ask, to raise a child when the mother will be in her seventies or even eighties before the child becomes an independent adult? We don't kill our orphans any more, but we do worry, perhaps with reason, who will care for them and whether aged parents, even if they survive, can properly bring up a small child.

Part of this worry is, of course, real and I share it too, but part of it is pure ageism. Many a child is born into a highly uncertain future even when its mother is young. And many a child has been raised very successfully by its grandparents. So why not older mothers? Women (and men) are living longer than ever before, and in better health. For many women today, the cruel words 'too late' come much too soon.

CHAPTER 12

Eat less, live longer?

Subdue your appetites my dears, and you have
conquered human nature.

Charles Dickens, *Nicholas Nickleby*

You are what you eat. If you add what goes in and subtract what
comes out, the answer is: you. All of your chemical and mineral
constituents, with the minor exception of any that remain of those
you got through your umbilical cord, got into your body by way of
your mouth. So it is hardly surprising that those of us who would
like to live a long and healthy life should pay careful attention to
what we eat and drink.

Food is the raw material of life, but it is possible to have too
much of a good thing. If you were to tell a subsistence farmer in
Navrongo – or any other similarly impoverished region of the world
– that going hungry might make him live longer, the kindest
response you might hope to receive would be a pitying shake of the
head. And yet, since the mid-1930s it has been known that, if you
restrict the food intake of laboratory mice and rats, you can
lengthen their lives by as much as a year. A year may not seem
much to you and me, but if your life span is just 3 years to start
with, a year's extension is a significant addition. The equivalent for
a human being would be about 30 extra years of life.

The phenomenon of life extension through dietary restriction has
spawned a huge number of scientific experiments. Effects similar to

those seen in rodents have been claimed in species as wide ranging as worms, spiders, flies and fish. Studies in monkeys are presently under way. And the big question is: does it work for humans?

Most of the evidence about underfeeding in humans, it has to be said, is bad news. Malnutrition is extremely harmful. The spectre of famine stalks much of the third world, even as the grain mountains of the affluent first world continue to grow, milk lakes fill, and heaps of surplus fruits spoil and rot. Hunger stunts growth and cuts short lives. If you are a woman, hunger harms the babies you carry. It can even harm the next generation if your baby is a daughter. Remember that a woman's egg supply is formed in the first half of pregnancy. If a mother is starved, not only will her daughter's growth be affected, but also her fertility.

Babies deprived of proper nutrients in the womb suffer disadvantages not only early in life when they are generally frailer than well-nourished infants, but also later in life when they are more prone to develop health problems like high blood pressure, diabetes, stroke and coronary heart disease. David Barker, epidemiologist with the Medical Research Council at the University of Southampton, suggests that this is because the undernourished foetus protects the growing brain at the expense of other organs, which may suffer permanent alteration in their metabolism and growth. Barker and his colleagues have made extensive studies of the association between birth weight and late-life disease, benefiting from the meticulous care with which midwives in Hertfordshire, Preston and Sheffield during the 1920s kept detailed records of their little charges.

Evidence that food restriction in humans is not all bad comes from the Japanese island of Okinawa. Okinawans consume, on average, just 80 per cent of the calorie intake of the rest of the Japanese population. The Japanese are anyway long-lived, which many ascribe to a diet that is rich in fish oils, vegetables and soy products. (Certain dietary ingredients are clearly good for longevity, as we shall see in Chapter 15.) Okinawa has the highest proportion of centenarians in the world (185 per million), four times as many as the rest of Japan.

Dietary restriction, sometimes described as '*under*nutrition without *mal*nutrition', provides all of the essential nutrients, but with a much reduced total energy intake. For many years there was a discussion about whether it was reductions in proteins, fats or carbohydrates that were most important. But careful research has shown that actually it does not matter. As long as the essential nutrients are there, it makes no difference whether food is restricted by cutting down fats, carbohydrates or protein. It is the calories that count.

In mice and rats, dietary restriction produces the following effects. Reduction of energy intake by 30 to 50 per cent relative to the *ad libitum* fed animal – that is, one that eats all it wants – increases both the average and the maximum life span by around a third. The size of the life span extension varies from one strain of animal to another. The life span increase is also affected by the method that is used to restrict intake. When you think about it, it is not that easy to control each individual mouse's food intake precisely, especially if several animals are housed together in a cage. One method, for example, simply restricts the times of day that food is available. The calorie-restricted rodent is lighter and smaller than the *ad libitum* fed animal, typically by one-third to a half. The size reduction is especially marked if calorie restriction starts early in life, soon after the young are weaned from their mothers. Calorie restriction also affects fertility; restricted rodents usually stop breeding altogether, or if they do have young, they eat them. This sounds rather grisly, but it is a frequent occurrence for rodents to eat their litters, both in the wild and in captivity.[17]

Apart from these obvious differences, calorie-restricted animals appear in many respects to be healthier than their *ad libitum* fed cousins. They perform better in tests of stamina and endurance, they have reduced rates of developing many types of cancer and, in keeping with their extended life spans, many of the biomarkers of ageing change more slowly. Within the cells and tissues of their bodies, they show elevated levels of some of the key maintenance functions, such as heat-shock proteins that sort out aberrant and damaged proteins. They also show a reduced accumulation of oxidative damage caused by free radicals.

For 60 years, scientists have puzzled over this phenomenon of life extension through dietary restriction and asked what causes it and why. Answers are beginning to emerge, and these will be crucial in evaluating its potential relevance to humans.

One obvious idea is that it is not in fact the calorie-restricted diet that is abnormal, but the diet that offers an unlimited food supply. Wild mice do not, as a rule, live their lives with 24-hour access to food in limitless abundance, as caged laboratory animals can do.

Some species seem to regulate their appetites to their needs, but humans, dogs and probably mice do this rather poorly. Huge numbers of people in the affluent societies eat vastly more than they need. Overeating in humans causes many well-known health problems and perhaps dietary restriction just helps a rodent to resist that extra snack that does the damage.

Alert to this criticism, dietary restriction researchers have shown that, even though some laboratory animals undoubtedly overeat, calorie restriction is still effective when compared with animals that are fed a controlled diet that avoids obesity but does not produce restricted growth and fertility.

Another idea is that calorie restriction reduces the overall metabolic rate in the restricted animals, slowing the accumulation of the toxic by-products of the chemical reactions that support life. This idea harks back to an early notion called the rate-of-living theory – a term coined in the 1920s by Raymond Pearl, eminent biologist at Johns Hopkins University in Baltimore – which is also linked to the wear-and-tear ideas of Chapter 5.

The rate-of-living theory is responsible for the quaint but widely accepted suggestion that all animals have only so many heartbeats and no more. If you take the heart rate of a mouse and multiply it by its life span, you find approximately the same number as if you do the same thing for a human or a horse. The theory is appealing because it captures an observation that has some empirical truth: small mammals do indeed live fast and die young. But small birds have higher metabolic rates than small mammals, yet on the whole they live longer. The correlations – and the exceptions to them – that form the basis of the rate-of-living theory are more successfully explained by the disposable soma theory that we encountered in Chapter 6 than they are by supposing that animals age and die at different times because they use up a finite stock of vitality at different rates. You can live a long time even if you have a high rate of metabolism, provided that you invest enough in somatic maintenance and repair.

When it was first suggested that a reduced metabolic rate might underlie the life-extending effects of dietary restriction, the early data seemed to support this notion. The gross metabolic activity of a typical calorie-restricted mouse is indeed less than the gross metabolic activity of an *ad libitum* fed mouse. But physiologist Edward Masoro, at the University of Texas Health Science Centre in San Antonio, was quick to point out the flaw in this interpretation. Masoro demonstrated that metabolic rate per gram of body weight is not reduced and is often increased in the calorie-restricted animal. Gross metabolic rate goes down only because the animals are smaller. As far as the accumulation of metabolic damage goes, it is, of course, the amount of metabolism per gram that matters.

Just why a mouse or rat should put *more* effort into its metabolism when food is scarce seems at first sight paradoxical, but there is an evolutionary logic that explains it. Animals often need to cope with a variable food supply. The good times are interspersed with bad. Regular and irregular shortages of food are something that natural selection is likely to have responded to in shaping the way that an animal lives its life or, in scientific terms, its life history. Some animals deal with the regular lean times of winter by entering a state of torpor, or hibernation. An alternative strategy, which may be better suited to unpredictable interruptions in food supply, may be to evolve the necessary degree of physiological plasticity to be able to switch resources between reproduction and maintenance when food is scarce.

In effect, what natural selection might have done for the mouse is akin to adopting the following line of reasoning:

> Resources are unusually scarce. Therefore I cannot do what I would like to do, which is to reproduce and maintain myself at appropriate levels. These will be the optimal levels of maintenance and reproduction, as determined according to the general circumstances of the species' ecological niche (following the same arguments as used in Chapter 6 to describe the mobbits – the disposable soma theory). If I attempt to reproduce with insufficient energy reserves, I might well end up destroying both myself and also my progeny. Therefore, I will shut down my reproduction and put all of my effort into maintenance, so that in a while, when normal food supplies are restored, I can take up where I left off without having aged too much in the intervening period.

Of course, natural selection does not plan in this purposive manner, but the idea sketches a possible route through which natural selection might have acted on the genes that determine how the mouse deploys its energy resources.

This basic idea was suggested in the late 1980s by David Harrison and Jeffrey Archer at the Jackson Laboratories in Bar Harbor, Maine, and by Robin Holliday at the National Institute for Medical Research in London, and has been elaborated and extended by others since. One extension is the notion that animals in times of famine need to put themselves into a state of heightened physiological alertness because food shortage in nature is likely to lead to increased risk-taking behaviour as the animal forages more desperately, fights harder to gain control of food, and tries eating unfamiliar, potentially toxic, foodstuffs. All of this calls for an elevated level of maintenance to combat and handle all kinds of challenges and stresses.

Any evolutionary hypothesis runs the risk of becoming a Just So story, along the lines of Rudyard Kipling's delightful but wildly implausible tales, if we cannot be confident that the idea at least makes sense in the rigorous, quantitative way that the mathematical dimension of Darwinian natural selection requires.

To weigh the arguments for and against the evolutionary hypothesis just outlined, my colleague Daryl Shanley and I developed a mathematical model of the mouse life history to examine the idea. We used all the information we could find on the feeding, breeding and survival of mice when not in captivity to build a theoretical model of a 'virtual wild mouse'. We looked carefully at the ways in which dietary restriction affects mice and built these effects into our virtual mouse model, which by this stage had grown into a rather complex computer program. We then asked our program the following question: if we challenge the virtual mouse with periodic bouts of food shortage, say over a three-month time span (time measured in virtual mouse months), and allow the virtual mouse to choose for itself the best fraction of its energy to allocate to maintenance, what will it do?

What we found was deeply interesting. When there was lots of food available to the virtual mouse, it did just what the disposable soma theory predicts it should do. It invests sufficient resources in maintenance to get it through its natural expectation of life in the wild environment in good shape, but not more than this. The upshot is that its life span settles at around 3 years. But if the food supply drops, and it can no longer manage both to maintain itself and reproduce, it *increases* the effort it puts into maintenance to a higher level than in times of plenty. In other words, the virtual mouse does just what Harrison, Archer and Holliday suggested it might do.

The beauty of a virtual mouse is that you can carry out all kinds of 'What if?' experiments without worrying about the sorts of ethical problem that have to be considered in real animal experiments, and you can also ask what would happen if the evolutionary set-up had played out a little differently. For example, Shanley and I found in our model that it made quite a difference just how the litter size was linked with the energy put into reproduction.

Two possibilities we considered were whether the number of babies in a litter was in direct proportion to the amount of energy put into reproduction, or whether the mice first had to invest some energy as a kind of reproductive 'overhead' before the first of the babies could be produced. An analogous situation occurs in printing, where the run-on costs of printing another 100 copies is usually much less than the cost of the first 100 copies, which must include the typesetting and set-up charges. If there was a reproductive overhead, it became less worthwhile for the virtual mouse to attempt the making of small litters. This reinforced our prediction that, in hard times, the virtual mouse would shut off reproduction completely and increase its investment in maintenance in the hope of successfully sitting out the famine.

Now you might ask: if dietary restriction is a response to *temporary* food shortage, how come it works if you restrict the animal's food supply for the whole of its life? The answer is that the response can continue long term, even if it *evolved* only to deal with short-term emergencies. In nature, if a mouse were exposed to food shortage all of its life, then it would not have any offspring and its Darwinian fitness would be zero. This is why the evolutionary response, if indeed it was such, would be most likely to result from food shortages that were relatively short term. Once again, the virtual mouse comes into its own because we can ask the program: what if the famines are assumed to be longer? The answer is that, if the famines are long, nothing will help the starved animal and you do not expect to evolve a life span extension through dietary restriction. In this way, the virtual mouse can be used to probe the likely evolutionary background as well.

Some gerontologists are confident that what works for a mouse will also work for a human, and that dietary restriction holds the key to human life extension. Just a few, notably Roy Walford of the University of California at Los Angeles, are so convinced of this that they are prepared to experiment upon themselves. Others argue that life extension through dietary restriction is peculiar to small, short-lived species because these are the ones that most need this flexibility in their life histories. Our work with the virtual mouse model suggests that the plasticity in small animals may indeed be greater than in humans, but we do not yet have enough data on the effects of dietary restriction in species near enough on the evolutionary scale to be sure. When results finally emerge from the ongoing dietary restriction studies on monkeys, we will know more.

But let us suppose for a moment that dietary restriction can work in humans. What would we need to do to gain our longer lives, and would we do it? Calorie-restricted rodents show the greatest extension of life span when food is restricted early in life, soon after weaning. Such practice would be ill-advised and almost surely unacceptable in humans because it stunts growth, as many know to their cost, and can interfere with physical development and learning. The eating disorder anorexia nervosa, when it occurs in adolescents, delays or blocks reproductive maturation and bone development. Nevertheless, even when started only in adult mice, dietary restriction has a significant though lesser effect on life span. An appropriate time to begin such a regimen in humans might be age 18. Abrupt onset of dietary restriction in rodents has been shown less effective than a more gradual scaling down of food intake, so let us suppose that the same would make sense for us too.

The target level for a calorie-restricted human might be one-third less than the normal 'maintenance' diet. But some experimentation will be necessary to establish an effective and acceptable degree of reduction. Herein lies the obvious challenge.

If we consider a typical maintenance diet for a male office worker of 2,000 calories a day, reducing this to 1,400 calories a day will cause extreme difficulties for many. We will need to find effective ways to mask appetite and trick our bodies into feeling sated, even though we have consumed an energy intake so low that it will leave most of us ducking into the nearest food store for a hefty snack. Goodness knows, most of us eat far too much. We continue to do this, and to eat the wrong things, even though we know full well that it is bad for us to do so.

In spite of being hooked on ill-conceived and, for many of us, wholly unattainable ideals of slender bodily perfection – an ideal not shared in Navrongo, where traditionally a sylph-like figure in a woman is thought to augur ill for fecundity – we are tempted daily with high-calorie products like chocolate bars, cream cakes and french fries. Sugar, that curiously addictive and wholly unnecessary dietary substance, which was introduced to Europe only in the sixteenth century and brought untold misery by its role in fostering the Caribbean slave trade, rots our teeth and wrecks our proteins. We all know the perils of sugar, but how many of us can give it up entirely? It is just too seductive.

The only dieting tip I have ever found effective is this: eat what you like, but eat slowly so that your body has time to signal to the appetite centres of your brain that food has arrived. Then ask yourself at the appropriate time 'Do I really need this extra forkful (or spoonful) of food?' If the answer is no, stop eating. Simple, cheap and it works! But it is not easy. My own particular problem is that I have always been a fast eater. I blame this, perhaps unfairly, on childcare guru Truby King and his accursed notion of 'schedule feeding'. I am convinced that this made me so desperate for each feed that even now I have not learned to slow down.

The real difficulty with healthy eating, as with many of the things that we know are good for us but find hard to do, is that the benefits of a healthy old age, and maybe even of some extra years of life, seem very remote when we are young and exposed to the temptation of immediate gratification. By the time we are older and our eye lenses are clouded with cataracts, our muscles weak and so on, much of the harm has already been done. Any preventive measures we might take then will be less effective.

Even if calorie restriction proves not to work in humans, or if we lack the will power to do it, the study of this remarkable phenomenon will throw further light on the ageing process.

CHAPTER 13

Why do women live longer than men?

What lips my lips have kissed, and where, and why,
I have forgotten, and what arms have lain
Under my head till morning; but the rain
Is full of ghosts tonight, that tap and sigh
Upon the glass and listen for reply,
And in my heart there stirs a quiet pain
For unremembered lads that not again
Will turn to me at midnight with a cry.
Thus in the winter stands the lonely tree,
Nor knows its boughs more silent than before:
I cannot say what loves have come and gone,
I only know that summer sang in me
A little while, that in me sings no more.

Edna St Vincent Millay, 'Sonnet'

In some countries, it is an article of law that women and men are equal – equal, that is, in terms of status, worth, rights and obligations. Even in Navrongo, where traditional marriage customs still require a man to pay a substantial bride price in livestock, or the cash equivalent, to 'buy' a wife from her family, there are signs of impending change. T-shirts can now be seen around the town proclaiming the slogan 'Liberation by any means', accompanied by graphic pictures of women brandishing Kalashnikov assault rifles. But however egalitarian the law may be, when it comes to biology, men and women are not the same. We differ in our organs of procreation, we differ in our chromosomes, and we differ in the lengths of our lives.

On average, women live longer than men. This chapter examines what the basis of this difference may be, and on the way looks at one of the most intriguing problems in biology: namely, why bother with sex at all? We need to understand something of why and how the sex differences between men and women have evolved, if we are to tease out the delicate puzzle of just how much the difference in life span is innate, and how much it is the product of social and behavioural factors.

When we look around the world, we find that the longevity difference between the sexes varies, being about 6–8 years in western Europe, North America and Australasia, 3–5 years in much of Africa and Latin America, and zero or even negative in the Indian subcontinent. This suggests at once that social factors are important, and we saw in Chapter 3 how differences in the treatment of young girls and boys might explain the absence of any female advantage in expectation of life in India.

If it were just in our own species that the sexes differed in life span, we might be tempted to seek an explanation purely in the ways that men and women behave. After all, in most societies there are major behavioural differences that will be sure to have some effect on average life expectancy. Men are more combative, and thus prone to injury. Men tend to be more competitive, in and out of work, and thus arguably at greater risk of stress-related illness. Men are more frequently engaged in physically dangerous work, and thus at greater risk of accidents. Men in the past have smoked more cigarettes, and thus run greater risks of cancer and heart disease. But the idea that behaviour alone can explain the sex difference in longevity seems unlikely.

The animal kingdom abounds with species that also show sex differences in life span. Generally, but not universally, females live longer than males. Male hamsters, guinea pigs and wolves live at least as long as females. So why do women live longer than men?

As the maternally derived half of you – the egg – bumbled and tumbled its way down your mother's oviduct, life could have gone either way. Racing to meet it was a wriggling horde of sperm, thrashing their whip-like tails as hard as they could in the frenzy to be the first to find and fertilise the egg.

Roughly half of these sperm carried an X-chromosome, and if you are a female, then it was one of these that won the race. The others carried a Y-chromosome, and if you are a male, one of these was the winner. In fact, the front runners were a little more likely to be male-making sperm rather than female-making sperm, because Y-chromosomes are a lot smaller than the X-chromosomes, and so the male-making sperm travel more lightly. The speed difference is not great because sperm of both kinds carry one of each of the twenty-three non-sex-determining chromosomes and these make up most of the payload. The egg always has an X-chromosome and the twenty-three non-sex-determining chromosomes. Thus two Xs make a female, and an X and a Y make a male.

The fact that Y-bearing sperm travel a little faster means that somewhat more than half of conceptions result in male foetuses. About 51 per cent of newborn babies are male. More than this percentage of conceptions are male, but male foetuses are a more likely, for reasons unknown, to undergo spontaneous abortion than females. This differential mortality continues throughout life (except where it is countered by social factors). By the time old age is reached, women significantly outnumber men.

Before we move to the whys and wherefores of these differences, let us look a bit more closely at the sperm race and at what it tells us about the male germ-line. The sperm that won the race and succeeded in giving you its genes was a sperm in a billion. Well, not quite a billion, but close. The average number of sperm released during sexual intercourse is around a third of a billion. Out of this enormous number, only about 200 will ever find the egg. To prevent more than one sperm from fertilising the egg, the first one to arrive triggers a change in the structure of the egg coat that effectively blocks any subsequent arrivals from gaining entry. Once the winning sperm is in, the egg shuts and bolts the door. One sperm is all she wants.

We saw in Chapter 11 how carefully nurtured and selected the egg is. Many eggs are discarded, perhaps because they are not up to scratch. And vast numbers of sperm are squandered too. All of this wastage results in pretty stringent quality control of the male germ-line. A sperm has to be well made to stand any chance in the competition.

The selection forces that act to eliminate defective sperm and eggs will contribute to the mechanisms that make the germ-line immortal. If cells are going wrong, perhaps because their molecules are making mistakes, it seems rather a good idea to set up a system so that each new generation can be restarted from the pick of the cells that remain.

In the case of humans, having a germ-line and relying on sex for reproduction go hand in hand, but this is not universally true. There exist some species that look as though they ought to have sex, but have no truck with it at all, although they do have a germ-line. Whiptail lizards are a recently discovered and rather surprising example. These attractive little creatures, to be found in the south-western United States, Mexico and parts of South America, are entirely female. They produce their offspring by parthenogenesis: virgin birth. They make eggs just like other lizards, but their eggs develop all by themselves with never a sperm in sight.

Plenty of women regard the male sex as something of a mixed blessing, and from a purely Darwinian perspective there are some undeniable advantages in being asexual and reproducing like whiptail lizards. Firstly, if you have no sex, then you are not for ever mixing someone else's genes in with yours and taking pot luck on the outcome. If your genes are already a winning team, and whiptail lizards seem to be doing OK, then why break them up? Secondly, being unisexual is a big help when it comes to colonising new territory. Suppose a tornado picks you up and drops you on an uninhabited island. If you can do without sex, then you do not have to wait for a mate of the opposite sex to get there too. But the third and biggest advantage of all is that a population of parthenogenetic females can reproduce at twice the rate of a population in which half is male. This arithmetical drawback is known somewhat drily as the 'twofold disadvantage' of sex.

When we look at sex like this, it is really quite strange that sexual species are so common and asexual species so rare. Birds do it, bees do it, even educated fleas do it, as the old song goes. Even an organism as unsexy as the yeast we use for brewing and baking has sex, in spite of the fact that each yeast is just a single cell and does not, in the ordinary sense, have sexual organs. In yeast, instead of males and females you get what are called mating type A and mating type B. But it is sex all right, low key as it seems to us.

As we broaden our horizons from the familiar joys of human sex and sexuality, the world of sex becomes more and more arcane and mysterious. For decades a star cast of evolutionary biologists like John Maynard Smith, whose papers prompted me down the path to the disposable soma theory, George Williams, whose theory of ageing we encountered in Chapter 6, and Bill Hamilton, who gave us the intellectual tools to understand the evolution of social behaviour, have grappled with the enigma of sex. And experimentalists have been hard at work to test their ideas.

One important idea gets its name from the weird fantasy world that Alice explores in Lewis Carroll's *Adventures through the Looking Glass*. As soon as Alice arrives in this bewildering domain, she finds everyone running and running and she must run to keep up. When Alice asks the Queen of Hearts why everyone is running so hard, the Queen replies, 'Now, *here*, you see, it takes all the running *you* can do, to keep in the same place. If you want to get somewhere else, you must run at least twice as fast as that!'

The 'Red Queen hypothesis' in evolutionary biology suggests the same thing. Natural selection is a harsh law. However well adapted you might be today, there is a predator, competitor or parasite out there just waiting to evolve the means to eat, beat or live in you tomorrow. So it goes. The only way your genes can keep from becoming extinct is to keep evolving.

When it comes to this sort of evolutionary arms race, parasites have one major advantage over their hosts. They can adapt a whole lot faster, mainly because they produce a lot more progeny. Many parasites are masters of the adaptation game. The frightening spread of resistance to antibiotics, following their widespread and indiscriminate use for what are often trivial complaints, is proof that not only are parasites able to keep one step ahead when it comes to biological evolution, they can also outwit all of the skill and ingenuity of present-day medical science. In developed countries, antibiotics tend to be overused and all too often a course is abandoned half-way through. In developing countries, antibiotics are often misused through lack of education, poverty, or because the stocks that are available are old and ineffective. An incomplete or understrength course of antibiotic fails to kill the infecting bacteria completely. The weakest bugs are killed, but the strongest, which may have gained partial resistance, survive to breed ever more resistant strains. In a remote area near Navrongo, I saw a farmer with a badly swollen, septic foot. This unfortunate man had bought two capsules of antibiotic, which was all he could afford, broken open the capules, and simply scattered the powder on the infected wound where he had cut himself with his hoe. The spread of drug-resistant malaria, such as kills millions in developing countries each year and killed my friend on his return from Navrongo, is another disquieting example of the Red Queen at work.

When in the 1980s Bill Hamilton suggested that sex was an antiparasite defence, many were sceptical. But the idea makes good sense and has gained ground steadily. The important thing about sex is that it shuffles the genetic deck of cards. The downside, as we saw when we discussed whiptail lizards, is that if you have a strong hand of cards, then having to shuffle them each generation in the process called genetic recombination seems a shame. But if you have to keep pace with a changing environment, then shuffling your cards to see if you can draw an even better hand might be a very good thing to do. How else might you quickly bring together a good gene carried by person A with another good gene carried by person B?

Gene shuffling by way of sex not only helps keep up with the parasites, it is also a good way to speed up the trial and error of genetic novelty that you might need to get out of other potentially catastrophic fixes, such as a change in climate. If your world is changing fast enough, this advantage of sex might outweigh its twofold disadvantage.

Another school of thought about why sex happens is the idea that sex is an important way of weeding out mutations. In time, any gene can become mutated as a result of getting damaged or copied wrong. DNA repair is astonishingly good, as we saw in Chapter 8, but even so, some mutations are bound to happen. If you have no sex, then you reproduce clonally. All of your descendants are genetically identical except for mutations. Mutations will accumulate down the generations unless eliminated by natural selection. The problem is that many mutations have only small effects. It may be a long time before these kinds of mutation take enough of a toll on Darwinian fitness to get selected against. By then, the chances are that all of the individuals in the population will carry at least some mutations. None of them is as good as the original.

Not far from where I live, at Alderley Edge in Cheshire, is a mysterious hill about which an ancient legend is told. Beneath the hill, in a secret cavern protected by deep magic and watched over by the Wizard of Alderley, there lie in enchanted sleep 140 knights in silver armour, each with a milk-white mare at his side. The knights are there to protect the world against the evil forces of the spirit of darkness that slowly but surely will poison the hearts and minds of us all and gain a foothold that cannot be gainsaid. To guard against this mortal threat, the wizards of old chose a band of warriors pure in heart and to keep them from contamination bound them in their magic sleep, from where they could ride forth again in time of direst need.

Something of the same idea lies behind the mutational theory of sex. We cannot, of course, lock up our purest genomes in a secret cave and protect them by force of magic from mutation. But we do need a mechanism to prevent mutations gaining a foothold that cannot be gainsaid. Sex rides to the rescue like this. If you have two sexually reproducing individuals, each of which carries a certain load of harmful mutations, then reproduction offers the chance of randomly begetting offspring with fewer mutations than either of the parents. This is because, on the average, half of the mutations from each parent get transmitted to the child, but by chance the actual number may be more or less than this. The process is a little like putting two hands of cards together, shuffling them, and drawing a single new hand from the pile. Sometimes you get a better hand than either of the two hands you started with. In the same way, sex provides the opportunity for some offspring to be less mutationally loaded than either parent. There are some who suggest that the *primary* role of sex is to prevent 'mutational meltdown' of the population, and there is evidence in yeast, at least, that this is so.

Whichever of these explanations of sex is correct, and perhaps both are, they share the need to bring the sexual partners together, in order that the two sets of genes can be shuffled. Some species, like yeast, maintain strict equality in how they go about this. Mating type A and mating type B meet and mate as perfect equals. Others, like us, evolved *la différence* and all that goes with it.

Way back in the evolutionary history of sex there began a small asymmetry, from which grew a bigger asymmetry that lies at the heart of many of the biological differences between men and women, perhaps including life span. The seed from which the difference grew is 'anisogamy' – unequal gametes (gametes are the cells that bring the genes together; in our case, the sperm and the egg).

If you need two partners to meet, you can either have them both go looking for each other, or you can have one play the role of seeker, while the other one waits to be sought. As soon as you evolve even the faintest beginnings of this division of labour between the gametes of the two sexes, you have started something significant because for the first time it begins to matter whether you are male or female. At the level of the gametes, the male sex is the seeking sex, the female sex is the sought sex.

The male gamete must invest effort in its search process and speed becomes important. Gametes compete with each other, even though they come from the same individual. Because of recombination (gene shuffling), each gamete is genetically different from its companions; each is a different draw from the genetic deck of cards. If there is a card in that deck that makes a male gamete travel faster than the others, more of the next generation will be ones that carry that particular card. Once started, this kind of selection quickly becomes a runaway process, and it is therefore no surprise that sperm have evolved to become little more than DNA packages with big outboard motors.

As the male gamete evolves to become an increasingly stripped-down search-and-deliver device, the female gamete needs to take on a greater and greater share of the responsibility for looking after the product of her union with the male gamete, when it arrives knocking at her cell wall. The egg is primed ready to begin unfolding the genetic blueprint of the new organism, and she carries nutrients to get it successfully started on its first steps of life. In the case of a human egg, for example, it will be a little while before the embryo can implant in the wall of the uterus and begin to draw nutrients from its mother.

In the earliest stages of anisogamous sex, the extent of the division of labour was probably modest, but we have come a long way since then. Multicellularity means that the female gamete is no longer an isolated cell, but is surrounded by a female soma, all of which has in one way or another evolved to serve the key biological function of propagating the genes. The male soma likewise has evolved to propagate genes via the male germ-line. It can be a bit startling to think of our different bodies, male and female, as being the product of a small asymmetry in the gametes of our long-distant unicellular ancestors, but it makes perfect sense.

Male or female, we are all the same species, and we share most gene functions in common. We have twenty-three pairs of chromosomes, the autosomes, that are common to both men and women, but the genes that differentiate between the sexes are special. Just one set of chromosomes, the X- and the Y-chromosomes, determine our sex. Soon after fertilisation happens, genes on the sex chromosomes begin to take us down our divergent paths. In humans, as in other mammals, it appears that females are what might be called the 'default' sex. That is, it takes the positive action of genes on the Y-chromosome to make a potentially female body (remember we all have at least one X-chromosome) into a male body. These genes begin by sending out a signal for making testes. The germ-line gets sorted out early on in the development of the embryo and, if you form testes, you get locked into the developmental path to maleness. If you do not get a signal to form testes, your germ cells form ovaries and you become female. Like everything in life, there are rare perturbations that can affect sex determination, but usually it works as simply as this.

Once the testes form, it is the sex hormones that call the tune. Testes make testosterone, the primary male hormone. Testosterone has a lot to answer for, including the shorter lives of males. Females make the hormones that after puberty set in train the business of ovulation, fertilisation, implantation, gestation and lactation. There can be no denying that, when it comes to mammalian reproduction, females carry the brunt of the effort. And all of this massive division of labour can be traced back to the primordial beginninings of a difference between seeking and sought gametes.

Looked at like this, it seems perhaps a bit odd that females are the longer-lived sex. On the face of it, women put much more biological effort into reproduction than men. This should mean that they have fewer resources left for maintenance of the soma, and indeed in some ways a woman's body does plunder her soma for resources like calcium when she is carrying her child in her womb and suckling it at her breast. But this is to ignore the biological side of being a male soma and making all those seeking gametes, the sperm.

There is rather compelling evidence that the investments in reproduction are not quite as one-sided as they appear. Males in many species invest rather heavily in all kinds of adaptations and behaviours that are tied into reproduction. Often the male is bigger, sometimes he has special structures like the antlers of stags that do not come cheap, and he commonly does a lot of rather risky fighting. Making all that sperm is not a trivial expense either. When biologists have carefully measured the lifetime energy costs of reproduction, it has turned out that the sexes are often pretty equal.

The other big difference has to do with disposability. In a mammal, the male's contribution to child raising is often much less than the female's. This makes the female intrinsically much less disposable than the male. Without her the child will probably die, as we saw so dramatically in the context of evolution of the menopause. The pressure to invest in maintenance and durability of the female soma is likely to be greater than the corresponding pressure in a male. This is revealed rather dramatically in the marsupial mice that we met in Chapter 10. The males burn themselves out in a mating frenzy and all die rather quickly. The females, on the other hand, need to live long enough to suckle their young and may indeed live another whole year.

It is this difference in disposability that may best explain why women live longer than men, the biological effects being brought out about through the actions of the sex hormones, particularly testosterone. We saw in Chapter 10 how surgical removal of the glands that in semelparous species trigger death can extend survival. Perhaps it will come as no surprise to know that there is some evidence that human eunuchs (male castrates) live a little longer than their intact brethren. On the other hand, the data are pretty patchy. There is no sign of queues of men forming to 'take the cure'.

We will close this chapter by looking at one question about sex and life span that naturally springs to mind. If, as the disposable soma theory suggests, we age because we put limited effort into maintenance in order to preserve resources for reproduction, can we opt not to have children but to have extra years of life instead? The question is entirely reasonable and if we were fruitflies the answer would be yes. Linda Partridge, professor of biometry at University College London, and her colleagues have shown this very convincingly. Virgin males and females live longer.

For humans, however, the answer appears to be no. Interestingly, there *is* a tendency for below-average fertility to be associated with above-average longevity. This was found by epidemiologist Rudy Westendorp and me in a study published in *Nature* in 1998 of extensive records of the births, deaths and marriages of British aristocrats. However, there is no evidence that those who simply elect not to have children live longer.

CHAPTER 14

The Genie of the Genome

> Sometimes, in a foal's crest, you can see
> Some long-extinguished breeding. So in us,
> The high-rise people and the dispossessed,
> The telly idols, fat men in fast cars,
> Something sometimes reverts to the fine dangerous
>   strain
> Of Galahad the high prince, Lancelot the un-
>   defeated,
> Arthur the king.
>
> U.A. Fanthorpe, 'DNA'

'If you would live long, choose well thy parents.' So says the Old Testament of the scriptures, and the writer clearly knew a thing or two about ageing. In your quest for longevity, you might want your parents to be rich, because a privileged lifestyle brings benefits of good nutrition, comfortable living and – in some countries – access to the best education and health care. All of these will give you the chance to make the most of your biological potential.* You might also want your parents to be caring and fun, because love and attention can make the best of even the most meagre circumstances. You should certainly choose your parents to have the right genes.

A number of genealogical studies have looked to see if traditional

---

* A recent study in a cemetery in Glasgow showed that the largest tombstones, signifying money in the family, had on average the greatest ages at death.

wisdom is really correct about the inheritance of life span. The results are positive. Long-lived parents have children who, on the average, live longer than the children of short-lived parents. What has not been so easy to determine is exactly how much of this is due to the genes. Wealth, education, a positive attitude to adversity and to personal health, and many other life-prolonging attributes can be passed down through the generations as well as DNA. Lately, some of the best data on the subject have emerged from that classic tool of the population geneticist, the study of twins. Identical (monozygotic) twins are formed when an embryo splits in two and have all of their genes in common. Non-identical (dizygotic) twins are formed from separate eggs and sperm and are no more alike than ordinary siblings; they have just half of their genes in common. A study published in 1993 examined the records of Danish twins born between 1870 and 1880, and showed that identical twins do indeed have life spans more similar to each other than non-identical twins.

Inheritance of life span is not clear-cut like inheritance of blood group or eye colour, the kinds of genetic trait observed by Gregor Mendel in 1865 as he grew his celebrated peas in the gardens of the Augustinian monastery at Brno in Moravia. The inheritance of life span is not even as clear cut as the inheritance of height, which is partly genetic and partly determined by diet. Length of life is influenced by many things, including purely random factors like car accidents. It is not only in the developed world that speed kills; the road to Navrongo was strewn with wrecks. All of this means that, in a population of individuals, a part of the variation in life span will be due to genes, another part will be due to lifestyle and environment – things like nutrition, smoking and exercise – and a third part will be due to chance. Twin studies allow an estimate to be made of just how much of all of the variation in life span is due to the genes. The answer is about a quarter.

The good news, if you have parents who were shorter-lived than the average, is that there is plenty of scope for other factors to boost your own life span. The not-so-good news, if you have long-lived parents, is that you cannot take above-average longevity for granted. There is a well-known genetic principle called 'regression to the mean'. Regression to the mean is why the offspring of parents who are at one or other extreme of the bell curve for some heritable attribute tend to move back towards the middle. Many a child of a gifted parent has had cause to rue the law of regression to the mean, chiefly because society can be cruel in its expectations of them. Longevity is no exception to the law. In the case of Jeanne Calment, fate was particularly unkind. Neither her daughter nor her grandson survived even to age 40.

Even though genes do not programme your life span in the same way that they programme your blood group, the understanding of what it is that genes bring to the complex business of ageing is vitally important. Our body's machinery may be subject to all kinds of non-genetic influences, but the machinery itself is programmed by the genes. We need to understand what genes do for ageing if we are to understand the ageing process itself, and if we are some day to intervene in its workings to improve the later years of our lives.

Mention genetics nowadays, and immediately there springs to mind the idea of scientists in white coats tinkering with the genetic blueprint, and of gene therapy just around the corner. In the words of *Longevity*, a magazine published until recently that described itself as a practical guide to the art and science of staying young:

> many researchers believe that specific genes in the DNA strand are programmed to break down and cause at least some of the physical symptoms of ageing, if not the entire ageing process. If this is true and if we could locate the 'age' genes, we might be able to use cloning or gene splicing to repair or eliminate them.

Well, maybe, and maybe not. We can be pretty confident from all that we have seen so far in this book that our genes are not actually programmed to cause the breakdown of the body. Our genes are programmed for survival. What we need to understand is which genes are the most important for longevity and how they work, singly and in combination. We can think of these kinds of gene as 'longevity assurance genes'. They are genes that control our soma protection systems, like DNA repair and the antioxidant enzymes. There are lots of soma protection systems, so there are likely to be lots of longevity assurance genes.

Here is a simple experiment that you can do in order to get some idea of how the idea of longevity assurance genes fits in with what we know about the genetic contribution to life span. You can make this a thought experiment if you wish, but there is no substitute for the real thing.

First, get hold of some plastic drinking straws. The exact number is not critically important, but try for a minimum of fifty. It will help if your straws are multicoloured, or if you can mark them with coloured pens or tape. Let us suppose that you have ten distinguishable types, with roughly the same number of straws of each type. Second, get a pair of scissors. Cut a short amount off the ends of most of the straws, keeping just a few at full length. Vary the length of the section you remove up to a maximum of about 5 cm, so that you end up with a batch of straws that differ in length. Try to ensure some variation in length within each type. In other words, you will have perhaps five red straws all a little different, five blue straws all a little different, and so on.

Your batch of straws represents a collection of longevity assurance genes, the length of any individual straw being the period of longevity 'assured' by that particular gene. For example, red straws might represent a gene controlling the activity of a DNA repair enzyme. If a particular red straw is a long one, it means that the enzyme level is set high. This in turn means that DNA faults will be repaired more aggressively, and DNA damage will build up rather slowly. On the other hand, if a red straw is a short one, the enzyme level is low, and DNA faults will build up more quickly.

If we now look at the set of straws as a whole, the average length represents the level that is favoured under natural selection, according to the disposable soma theory. This average length will be neither so short that it results in premature death, nor so long that it results in energy being wasted on better maintenance than is necessary. Remember from the mobbits of Chapter 6 that the straws only need to be long enough for the soma to get through its natural expectation of life in the wild environment in good condition. On the whole, we expect the average lengths of the red straws to be the same as the average length of the blue straws, and so on, because, as we saw in Chapter 9, natural selection works a bit like Henry Ford and his motor car mechanics. It avoids weak links not only at the organ level, but also at the level of the soma protection systems.

Natural selection is a potent force, but not an exact one, so we should not be surprised to find some small variations within the population in the levels of individual soma protection systems. In you, DNA repair might be a bit better than in me, but your antioxidant enzyme levels might be a little less than mine, and so on. This is reflected in the variation in the lengths of the straws that you have just produced with your scissors.

Now, let us suppose that the ten distinguishable types of straw represent ten different longevity assurance genes in the organism. There may well be a lot more than this, but ten is enough for the purpose of illustration. Put all of your straws in a small cardboard box or a bag, and pick straws one by one at random until you have one of each type. Once you have picked one red straw, put any further red straws back in the box, and so on. This represents the genetic endowment of one individual. Now it may be that the straws you have picked are all on the long side, in which case the individual is robustly endowed with longevity assurance genes, and can expect a good long life. Or it may be that the straws are all on the short side, when the reverse is true: you have literally drawn the short straws of life. Most of the time, you will find an average result, with some long straws, some shorter ones and some in the middle. You can do this over and over again and each time you will get a different result.

This first part of our experiment has introduced the idea of genetic variation in longevity assurance within the population. Next we tackle the problem of inheritance. For this we draw two lots of ten straws as above, each lot to represent a parent. Put only these twenty parental straws in the box, and then draw a set of ten straws from them to represent the child.

We can now repeat this whole process as often as we wish and observe the patterns that we get. In most of the draws, both of the parents, and consequently most of the children, will be fairly average. However, every now and then one or both parents will get unusually long (or short) sets of straws, and the child's straws will also be longer (or shorter) than the average. If only one parent gets an unusual set of straws and the other parent is average, then the child's set of straws is likely to be unexceptional.

This experiment captures the essence of the genetic component to life span. In real life the situation is, of course, more complicated. Not only does each individual have duplicate chromosomes, and hence two randomly drawn straws for each longevity assurance gene, but some genes are dominant whereas others are recessive. With patience we could elaborate the straw-drawing game to take account of such features, but it would become a lot more complicated. Rather than take the straw game further, let us now look at how some of these ideas can be studied using living organisms.

The organism of choice for many population geneticists is the fruitfly. Fruitflies have been used for genetic studies of many different kinds for nearly a century, not least because they are cheap and easy to raise. Recently, fruitflies have been used to probe the genetics of life span.

You can do with fruitflies what would be unthinkable with humans: namely, breed the little creatures selectively to try and produce an altered life span. The most interesting thing to do is to select for increased life span because all kinds of mutations can shorten life which have little or nothing to do with ageing.

But there is an unusual problem in breeding for life span, which does not arise when you breed for other traits like the number of bristles on a fly's body. Should you want to breed flies with more bristles, you simply pick the flies with the most bristles, mate them, then in the next generation pick the bristliest ones again, mate them, and so on. Provided that there exists a genetic variation in bristle number, you will end up with flies carrying a greater average number of bristles.

The problem with life span is that, by the time you know how long a fly has lived, it is dead and not much use for breeding! One way around this problem was devised in the early 1980s by geneticist Michael Rose, then at the University of Sussex and now at the University of California at Irvine. The trick was not to wait until the flies dropped dead, but to breed from them as old as possible – in effect, to breed for the ability to stay fertile into old age. The idea was that flies which retained the most fertility in late life were ageing more slowly than the rest of the population. By selecting for late reproductive ability, maybe they could select, indirectly, for increased life span. And it worked.

Not only did the flies live longer and prove to be more resistant to stress, but it also turned out that the long-lived flies paid for having more durable somas by trimming their investments in reproduction early in life. The young flies in the long-lived stocks laid fewer eggs, a result that has now been investigated rather comprehensively by several other groups around the world. There is some evidence that women who have lived past 100 retained their fertility somewhat later than the general population, possibly with a later menopause. There is also evidence that among the English aristocracy, for whom reliable records extend across several centuries, women with greater longevity had lower overall levels of fertility. These data are consistent with the fruitfly findings.

Another organism that has become a firm favourite of geneticists interested in life span is a curious little worm that lives in the soil. This is not the common earthworm, but something much smaller; a nematode so small, in fact, that you need a microscope to see it properly. When you do see it through the microscope, it is a shiny and graceful creature. It goes by the Latin name *Caenorhabditis elegans*, or *C. elegans* for short.

Biologists became interested in *C. elegans* in the 1970s when it was found that these tiny little creatures have a rather precise growth process. Each adult has exactly 959 somatic cells in its body, no more and no less, unless it is a mutant. This is a far cry indeed from the hundred thousand billion cells that you and I have in our bodies, and it held out the prospect of understanding *exactly* how the nematode forms its body and lives its life, a prospect that is well on the way to being realised.

What biologists have also found is that there are some mutant nematodes that live longer than usual. The genes in which these mutations occur are beginning to be understood. Many of the mutations do something rather interesting because they increase the nematode's ability to cope with stress – for example, the stress of temperature. In other words, they make the soma more durable. We now know that one of the genes is involved in regulating energy metabolism, which should not surprise us at all.

We have seen that nature provides for genetic variation in fruitfly life spans that can be acted upon by artificial selection in the laboratory, and that mutations can be found in nematodes that increase life span. But can an increased life span be manufactured by genetic engineering? In fruitflies, nematodes and mice, scientists are now trying hard to make genetically altered animal models to see whether it will be possible to make animals live longer by introducing extra copies of longevity assurance genes, like the genes that code for antioxidant enzymes or for heat-shock proteins. In terms of our earlier experiment with the straws, the aim is to make the straws a little longer!

We will look later in this chapter at what this research means for the future of human longevity, but before we do this, we will also consider another kind of gene that affects the life span. These are not genes for longevity assurance, but genes that affect the risk of disease.

There are some 200,000 genes in the human genome and most of us carry mutations in at least some of our genes. Often a mutation is silent because the mutant gene on one chromosome is masked by a normal gene on the other chromosome. Such mutations are called recessive. We saw in Chapter 10 how the retinoblastoma gene, a tumour suppressor gene, is an example of a gene whose mutations are recessive. Another example is the cystic fibrosis gene. Mutations of the cystic fibrosis gene are carried on one of their two chromosomes by about one in twenty of most European populations, making cystic fibrosis one of the commonest genetic disorders. Children whose parents are both carriers of mutant genes have a one in four chance of getting two mutated gene copies, leading in time to the accumulation of thick mucus in the lungs, severe problems with digestion, and a cruelly shortened expectation of life.

When the cystic fibrosis gene was discovered in 1989, it was thought that not only would it speed the discovery of a cure, but also, while the cure was being discovered, prospective parents could be screened for the mutation and advised accordingly. If only things were so simple! Quite apart from the ethical difficulties that arise when genetic tests become available, the cystic fibrosis gene has turned out to have so many different mutations that doctors have a tough time in deciding which of the variants they should test for. When parents already have one child with cystic fibrosis, the variants they carry can be identified and future children can be screened, providing reassurance when the child is free from the disease or the choice of whether or not to abort an affected foetus. But mass screening of the general population is another matter altogether, both ethically and in terms of scientific reliability.

A few mutations in the human genome are not recessive but dominant. They affect you even if only one chromosome carries a mutated copy of the gene. Dominant mutations are much rarer than recessives because they mostly get eliminated rather quickly through natural selection. If a dominant gene acts early in life, it will not even be reproduced, so each case is a new mutation, instead of one that has been passed down through the generations.

But there are some dominant mutations that wreak their havoc only later in life. The mutation that causes Huntington's disease is a classic example.

Huntington's disease does not became apparent until middle age, by which time the sufferer may well have had children, half of whom can expect to carry the mutation and will in time get the disease too. The disease is unusual not only in the fact that it is carried by a dominant mutation, but also because the mutation takes a rather odd form. Most mutations either change a little bit of the DNA sequence, causing the protein made from the gene to be altered in its amino acid sequence, or delete a whole chunk of the DNA, even eliminating the gene altogether. What happens in the Huntington's mutation, however, is that a triplet in the DNA sequence gets repeated over and over again, as when someone stammers.

Because each triplet in the DNA sequence gets translated into just one amino acid in the protein sequence, as we saw in Chapter 8, the stammer gets translated as a string consisting of the same amino acid, over and over again. The odd thing is that the normal gene also stammers, but it does so a limited number of times. Anything between ten and thirty-five repeats can be found in unaffected individuals. But when the stammer mutates to exceed forty repeats, disaster strikes.

Nothing shows wrong for forty or more years, but eventually the elongated form of the protein results in the death of certain neurones in the brain, causing symptoms that include involuntary movements, loss of balance and co-ordination, depression and changes of personality. It is a harrowing prospect, made in some ways all the more agonising now that the gene has been characterised. If one is the child of an affected parent, one can take a genetic test with a 50 per cent chance of knowing that one is completely in the clear, and a 50 per cent chance of having one's fears confirmed.

Cystic fibrosis and Huntington's disease are instances where having a certain mutation, either as two copies in the case of a recessive condition, or one copy in the case of a dominant condition, actually causes the disease. There are many cases, however, where genes appear not so much to *cause* diseases directly, as to *predispose* towards them. There are genes that predispose towards cancers, such as breast cancer and colon cancer. There are genes that predispose towards heart disease. There are genes that predispose towards arthritis. And there are genes that predispose towards Alzheimer's disease. The difference is that a gene which predisposes you to a particular disease increases the likelihood that you will get the disease, but it is by no means a certainty. It may be that the gene makes it possible that you will suffer earlier onset of a disease that is common among the elderly, such as arthritis, or it may be that the predisposing gene requires some other factor, known or unknown, to trigger its effects.

In some cases, the genes that predispose towards disease are variant forms of normal genes that occur commonly enough in the population that they are not regarded as mutations at all, but as alleles. Many genes are polymorphic – that is, they exist in the population in various alternatives, just like different brands of beer. These alternative forms of the gene are its alleles. One of the better-known cases of polymorphism is involved in tissue matching. There are several genes that determine tissue type, and each of these has multiple alleles. The number of possible combinations is immense, which is why finding a compatible donor for a bone marrow transplant can be so very difficult.

The case of Alzheimer's disease shows just how complicated it is to unravel the genes that predispose towards common diseases. The first pointer to the role that genes might play in Alzheimer's came from the observation that people with Down's syndrome were found at autopsy to show the widespread plaques and tangles in the brain that are used to confirm a diagnosis of Alzheimer's disease, even though people with Down's syndrome have a much shorter life expectancy than the general population, generally not living beyond their forties. The early onset of these symptoms in the brains of Down's syndrome individuals suggested that there might be a genetic factor that somehow accelerated brain ageing.

Down's syndrome is caused by having one too many copies of chromosome 21 (three instead of two), so it was natural to look for a gene on chromosome 21 that might be involved in causing the brain lesions. The problem in Down's syndrome is not mutation, but something called gene dosage. If you have three copies of a gene instead of two, you often get too much of the protein made by the gene – not always, however, because some gene products are carefully controlled in the cell and production is strictly regulated according to need.[18] It happens that one of the genes on chromosome 21 is a gene that makes the amyloid protein, an abnormal form of which – beta-amyloid – is found in the nerve cell plaques in Alzheimer's disease.

Another clue to the genetics of Alzheimer's disease is that there are some families with a history of early onset, in which the disease begins in a person's forties or fifties instead of their sixties, seventies or later. In the early 1990s it was found that some of these families carried *mutations* in their chromosome 21 genes that made the amyloid protein.

Curiouser and curiouser. Three copies instead of two as in Down's syndrome, or a mutated gene as in the familial case, and in all cases the abnormal beta-amyloid protein is found at the scene of the crime! In good Agatha Christie style, the evidence was beginning to stack up against beta-amyloid.

But within just a year or two, it was found that only in about twenty or thirty families worldwide with history of early onset Alzheimer's disease was the amyloid gene mutated. In more than a hundred other families, also with early onset Alzheimer's disease, the amyloid gene was perfectly intact. It turned out that another gene, which was at once named presenilin, was involved in most of these other families. And then, in 1995, a third gene was discovered, and so presenilin became presenilin 1, and the new gene was named presenilin 2. All of this progress has been very exciting and it holds out the promise that we will understand a lot more about what causes Alzheimer's disease in due course, but the familial cases are only a tiny fraction of all of the cases of Alzheimer's disease, so what about the rest?

In 1993, it was discovered that a gene called apolipoprotein E might be implicated in the more usual non-familial cases of Alzheimer's disease. Apolipoprotein E is one of a family of genes that have been studied mainly for their possible involvement in cardiovascular disease. Apolipoprotein E has three common forms, or alleles. These are known as epsilon 2, epsilon 3 and epsilon 4. Each of us has a pair of these alleles, so I might be 4/4 whereas you might be 2/3, or any of the other possible permutations. But I rather hope that I am not 4/4 because the epsilon 4 allele is associated not only with elevated risk of heart disease, but also with elevated risk of Alzheimer's disease.

It is important to understand that for all the buzz and excitement that accompanies these genetic advances – and make no mistake, they take us a big step forward – we still have little idea of the role played by beta-amyloid in causing Alzheimer's disease, or how presenilins 1 and 2 fit into the story, or what the apoliprotein polymorphism signifies. The genetics of cystic fibrosis are at least conceptually straightforward. A gene is broken and needs fixing. The genetics of diseases like Alzheimer's, and the ways we can use the understanding that will come from unravelling them, will be far more complicated.

A part of the complexity is that, when it comes to gene polymorphisms, it makes little sense to think in terms of broken genes or even, perhaps, bad alleles. I said that I hoped I do not have two copies of the apolipoprotein epsilon 4 allele, but this is based on our very limited information about how this allele really differs from epsilon 2 and epsilon 3. The epsilon 4 allele is common enough in the population that its presence has almost surely been maintained by natural selection, and this means that it cannot be all bad news. However, we cannot be sure because times may have changed.

Around Navrongo and in peoples worldwide of West African origin, there is an allele of the blood protein haemoglobin that causes tremendous misery and suffering. This is the sickle cell variant of haemoglobin. People who are homozygous for this variant – that is, both their gene copies are of this form – suffer from the severe and potentially life-threatening episodes of sickle cell disease, when their red blood cells alter shape and cannot pass so readily through the small blood vessels. But people who are hemizygous – that is, they have one allele of sickle haemoglobin and one allele of normal haemoglobin – are partly protected against the lethal effects of malaria, still a common killer in many regions. The problem is that you cannot have the protection for heterozygous sickle cell carriers without the risk that two heterozygous carriers will have homozygous children. If the sickle cell allele is too common, too many people get sickle cell disease. If the sickle cell allele is too rare, too many people run the risk of dying of malaria.

We understand the good and bad side of sickle haemoglobin quite well, and in malaria-free regions of the world, we know that sickle haemoglobin causes only problems, but we do not yet know the nature of the trade-offs for polymorphic genes like apolipoprotein. It may be that the epsilon 4 allele is good in ways as yet undiscovered, which still offset the late-life disadvantages of heart disease and Alzheimer's disease. Or it may be that the epsilon 4 allele was good under the conditions of our Stone Age ancestors' lives and, like sickle haemoglobin in developed countries, is no good any more. We need to know more before we make hasty judgements.

One intriguing approach to studying the effects of gene polymorphisms in ageing has been pioneered by geneticist François Schächter at the Centre for Study of Human Polymorphism in Paris. Schächter and his colleagues gathered blood samples from more than 300 French centenarians and compared them with samples from younger individuals from the same populations to see if there was any evidence for genetic differences that might yield clues to longer life. This is an approach that has potentially a lot to offer, and a few years ago Schächter, Daniel Cohen and I suggested how it could be used to investigate the kind of idea that we encountered earlier in this chapter in the straw-drawing experiment. Do centenarians have longer straws than the rest of us – for example, do they have better DNA repair?

So far, however, the approach has mostly been used to study polymorphisms like apolipoprotein E. And it has turned up some interesting results already. In the case of apolipoprotein E, perhaps not surprisingly, fewer of the centenarians carried the epsilon 4 allele than in the general population. This suggests that those at higher risk of heart disease and Alzheimer's disease are less likely to live past 100. But another gene, called angiotensin-converting enzyme, or ACE for short, gave a surprising result. An allele of ACE that is linked with risk of heart disease was *more* common among the centenarians than among the general population. This suggests that the allele may confer some as yet unknown benefit, perhaps at older ages, that can outweigh the disadvantage of the increased risk of heart disease in a person's fifties, sixties and seventies.

The very oldest people are remarkable in many ways. Not only have they outlived most of their peers, but they are often fitter and healthier than people 20 or more years younger. Their genetic endowments and the ways they have lived their lives have many secrets to reveal.

The understanding of human genes is growing fast, but with it comes a growing worry that the Genie of the Genome will prove as mixed a blessing as any genie of legend. Many a fable points up all too clearly the folly of man when faced with the kind of power that the human genome project may reveal. Think of Midas, with his golden touch, who ended up by destroying what was most dear to him. Already we see the beginnings of a genetic underclass emerging. Anyone who tests positive for the Huntington's gene, for example, will have an extremely difficult fight to get any life insurance. While no insurance company can be expected to take such an individual on at ordinary rates indefinitely, it should be stressed that Huntington's disease does not strike until middle age. Surely a policy could be issued to provide life cover to age 35, or even just a policy that excludes death from Huntington's disease and its possible complications.

The whole principle on which insurance is based is to share and spread risk and, if we can now quantify risk ever more precisely, then we need to devise correspondingly intelligent ways to ensure that those who are disadvantaged in their genes are not doubly disadvantaged by society's failure to protect them financially.

The other big fear concerns privacy, confidentiality and simply coping with genetic information. Will we be required to have genetic tests before employers hire us? Do we actually *want* to know what our genes say about our future health and longevity?

In the long run, genetic knowledge has enormous potential to let us make the most of our lives. It need only be a Pandora's box if we allow it to be, and the surest way for that to happen is to let others write the rules for us. Ignorance and misconception about what modern genetics can and cannot do are all around us. Even scientists talk about advances like the discovery of presenilin as the gene *for* Alzheimer's, as if one gene alone caused this complex disorder. Geneticists who tout their advances as heralding 200-year human life spans coming soon, or who condone journalists who hint at this, are contemptible not only for their disregard of scientific veracity, but also for duping the public who pay for their work.

It is my deep conviction that science should not be done in a vacuum and that all of us should share in writing the rules by which scientific insights, particularly genomic insights, will be exploited. It behoves scientists to communicate their advances to the public, and it behoves the public to press scientists to do so, to pay attention and to uncover obfuscation. Yes, we are understanding the genetic basis of life span better and better each year, and yes, let us hope that this knowledge will uncover ways to improve the quality of life in old age. Let us approach the Genie of the Genome with humility, determination and hope.

CHAPTER 15

In search of Wonka-Vite

Sometimes our best efforts do not go
amiss; sometimes we do as we meant to.
The sun will sometimes melt a field of sorrow
that seemed hard frozen: may it happen for you.

Sheenagh Pugh, 'Sometimes'

The label on the bottle read as follows: 'WONKA-VITE. Each pill will make you YOUNGER by exactly 20 years. Caution! Do not take more than the amount recommended by Mr Wonka.' But of course, what did Grandma Georgina do? She took four pills all at once. Which might have been all right if, like Grandma Josephine, she had been 80 years and 3 months old. But she wasn't! Grandma Georgina was just 78 years old. And so she entered Minusland, from where Charlie and Mr Wonka, by dint of their great courage and perspicacity, rescued her wraith-like presence with four squirts of Vita-Wonk.

Vita-Wonk, of course, does the opposite to Wonka-Vite. It makes you older. The tale is from Roald Dahl's delightful book *Charlie and the Great Glass Elevator*. Read it if you haven't already, preferably out loud to a 5–8-year-old.

There is plenty of Vita-Wonk in the world, to be smoked, drunk, eaten and otherwise enjoyed or abused. But is there any Wonka-Vite? Can it be made? The ancients thought there was, and that it could. The not-so-ancients did as well, and now many moderns do too. At least they very much hope so.

As a gerontologist, a question I am often asked is 'What are *you* taking?' A few years ago at a conference on ageing in Italy, the manufacturer of some purported anti-ageing 'nutritional supplements', mainly trace elements like zinc, thoughtfully donated several cartons of free samples. They were gone in a flash! Delegates were to be seen staggering back to their hotel rooms, clutching armfuls of the stuff and coming back for more. I did too. But did this mean that we believed in it? Probably not. Mostly I suspect that we were victims of 'freebie syndrome'. It is extraordinary what can be got rid of, if it is offered as a free sample.

I am not, of course, being entirely serious. Vitamins and mineral supplements are very good for those whose diets are deficient in them. Many older people have diets that could benefit from this kind of supplementation. But supplements should be used with caution. A car needs oil in its engine, but it does it no good and even some harm to give it too much. The fact that a deficiency is bad for you does not mean that a surplus will do you good. Most of my free samples stayed behind in my hotel room in Italy.

My own anti-ageing measures are unremarkable and consist mainly in the avoidance of Vita-Wonk. I do not smoke, although I confess to quite liking the smell of freshly lit tobacco (stale cigarette smoke is another thing altogether). I avoid fatty foods. I eat fresh fruit and vegetables in preference to most meats. I avoid sitting for long periods in front of the television and I remind myself to go running from time to time, but not as often as I should. I try also, as far as is possible, to avoid undue stress. Something I find stressful is driving in rush-hour traffic, so last year I sold my car and I now travel to work by train. It means an hour's walking each day, but I feel the benefit of the exercise, and I can catch up on my scientific reading as the train rattles down the Tame Valley and into Manchester. It also means one car fewer on the roads and a few thousand cubic metres less exhaust gas each year. The sources of Wonka-Vite that I prefer are good fresh food, preferably vegetables grown organically in our garden, fish, red wine, exercise, and the stimulus of family, friends, pets, music and books.

A number of recent epidemiological studies have found that vegetarians tend to live longer, healthier lives than meat-eaters. Just what causes the difference is unknown. It might be that eating meat actually does something to shorten life, or it could be that the vegetarian diet contains something good for longevity, such as tofu, of which meat-eaters do not get enough. The Japanese diet is high in vegetables and soya products, which is thought by some to explain why the Japanese live 3 years longer, on the average, than the Americans and British.

Fish is another major ingredient in Japanese food that is good for longevity. Research has shown that fish oils may have protective effects against heart disease and stroke. Fish oils are rich in omega-3 fatty acids that, compared with the saturated fats found in meats, do not harden as readily and may stick less to artery walls. Many years ago, I was involved in a clinical study of haemophilia in which volunteers would come into the laboratory each morning, roll up their sleeves and give a syringe-full of blood. The blood would be spun hard in a centrifuge to separate the cells from the clear straw-coloured plasma. One young man's plasma was as cloudy as milk. The cloudiness was explained when he was asked what he had had for breakfast. A fat-laden breakfast of cheese, sausage and fried eggs had filled his blood stream with the fine suspension of fats that had made his plasma so opaque. It was an illustration of how fat enters the blood stream that has remained vivid in my mind ever since.

Sanatogen Tonic Wine has been sold in the UK for as long as I can remember, long before the habit for wine drinking spread much beyond the upper classes. Nowadays wine comes from almost every region of the globe, and can be bought in almost every supermarket. Great glee, therefore, accompanied the discovery a few years ago that Sanatogen were right and that a glass or two of wine a day is good for you. Other kinds of alcohol, taken in moderation, are thought to be beneficial too.

But beware. Alcohol is a case where Wonka-Vite turns to Vita-Wonk if taken in excess. Russian life expectancy for men has taken a significant tumble since the 1980s, from 64.9 years in 1987 down to 57.6 years in 1994. The likeliest culprit is thought to be the increase in alcohol consumption that has occurred over the same period.

Variety is the spice of life and the one occasion when I break all my usual dietary rules is when I breakfast in the United States. American breakfasts are too tempting; fried eggs over easy, bacon, hash brown potatoes, pancakes and maple syrup. I love it, never mind what my plasma looks like afterwards!

So when I attended my first American conference on ageing in 1978, I was enjoying a veritable cholesterol feast one morning, when to my surprise and consternation a pleasant young couple joined me and sat down to munch their way through a bowlful each of chemicals. The couple were Durk Pearson and Sandy Shaw, whose soon-to-be-published bestseller *Life Extension: A Practical Approach* was to give the world a spatula-by-spatula guide to concocting a breakfast just like theirs. The recipe for their chemical cocktail may be found in Appendix G of their book and comprises a powder mixture of assorted vitamins (A, B, C, E), L-arginine, L-cysteine, rutin, zinc, selenium, hesperidin complex, dilauryl thiodiproprionate and thiodiprionic acid, with various other tablets and substances taken separately (including antacids, as required).

I can manage oatmeal every day, but I draw the line at chemicals. A case can be constructed for Pearson and Shaw's dietary recommendations, which is well expounded in their book, but there is as yet little evidence for the efficacy of most nutritional supplements. This, of course, is why they are sold as such, and not as pharmacologically active substances.

Antioxidants are actively promoted nowadays as agents to help combat ageing. It makes some sense to suppose that, if you add antioxidants to your diet, you may stave off some of the ravages of free radicals. But remember that your cells already know about the dangers of free radicals, and have done so through much of their evolutionary history. Compared to the potent antioxidants that your cells manufacture already, the antioxidants you take on a spoon or in a tablet may make little difference, even if they reach your cells in a biologically active form. And it is as well to beware of tinkering with what we do not yet fully understand. One antioxidant on offer is an enzyme called superoxide-dismutase, or SOD. Having too much SOD is a mixed blessing antioxidant-wise. More SOD means that superoxide radicals get converted to hydrogen peroxide more quickly. But you need to do something to the hydrogen peroxide too, for it is also a free radical. Inside your cells, SOD works as a part of a team with other enzymes like catalase, which renders hydrogen peroxide harmless by converting it to water. Changing just one component in the finely balanced antioxidant system may not be a good idea, as we see in the case of Down's syndrome. In individuals with Down's syndrome – caused by having three copies of chromosome 21 instead of the usual two – the extra chromosome 21 carries an additional copy of the SOD gene. There is evidence that too much SOD is responsible for some of the trouble.

Having issued this cautionary note, I do believe that some nutritional supplements are worth taking, provided that informed medical opinion confirms that they are not toxic. Vitamin C is well tolerated by the body in high doses, although above 4 g a day it has been linked to kidney stone formation and gout. As we saw in Chapter 8, vitamin C provides a very general kind of antioxidant protection, acting as a recyclable punch-bag for free radicals.

Vitamin E supplementation makes sense in the same kind of way, because it quenches free radical chain reactions in membranes. Two American studies published in 1997 described different benefits from vitamin E. A multicentre team led by researchers at Tufts University found that a daily dose of 200 mg of vitamin E can improve immune function in older people, while another multicentre team showed that even higher daily doses of 2,000 mg appeared to delay slightly the progression of mid-stage Alzheimer's disease. The recommended dietary allowance of vitamin E is 30 mgs daily. (It has to be said, though, that other trials of the purported benefits of vitamin E have turned out negative.)

There may also be a case for modest doses of trace elements like selenium, which our bodies need in tiny amounts. Selenium has various roles to play in the body's antioxidant defences. We probably get fewer trace elements nowadays with our squeaky-clean supermarket vegetables than we did in times past.

Even if nutritional supplements did you no good at all, there is still the placebo effect. This can make you feel wonderful from nothing at all. Once I visited a delightful temple in Kyoto where water gushed from three fountains. The water from the first fountain was reputed to make you rich. The water from the second brought you love and happiness. The water from the third gave you long life. Well, of course, I took my cup, drank from all three and felt at once brimful of confidence for the future.

The placebo effect should not be underrated and it opens an important window on to the effects of the mind on the body. Once rubbished by medical science, the idea that mind can influence the state of the immune system is now accepted. Chemicals secreted from the brain undoubtedly modulate some aspects of the immune system. Psychoneuroimmunology – the study of how the brain and immune system interact – might be a mouthful, but it has become a respectable, even exciting field of research.

There are some who believe that mind-power alone can stave off the ravages of time. The Eternal Flame foundation is one such example. There is little to say scientifically about these claims, except that there is simply no evidence to back them up. In the case of those who support such claims with pseudoscientific arguments, the matter is more disturbing. Medical and scientific knowledge blended together with mysticism and belief can form a heady confection as appealing and insubstantial as a rainbow. If you are good with words, and if those you address lack the grounding in science that is needed to judge matters for themselves, this may exert a powerful draw.

Science does not have all the answers, but the successes of

science *are* tangible; they can be checked. Mind-power has had no comparable successes, and its failures are presumably put down to simple lack of faith, and faith cannot be measured. I would be inclined to dismiss the advocates of mind-power entirely were it not for the placebo effect. What such advocates give their followers is hope, and the placebo effect presumably springs ultimately from hope. But placebo effects are limited and the real hope of progress in combating ageing is in science.

Just before we get into the current and future science, we must acknowledge that the historical catalogue of scientific and pseudo-scientific attempts to find Wonka-Vite includes some entertaining fiascos. Metchnikoff with his Bulgarian bacillus (see Chapter 6) was soon followed in the 1910s and 1920s by a spate of bizarre claims in America and Europe that testicular transplants not only could revitalise flagging sexual prowess, but would in time abolish or at least postpone old age. In the United States, gland grafting was championed most spectacularly by 'Doc' John Romulus Brinkley from North Carolina, a man whose medical training was rudimentary, but whose public relations skills were superb and whose clients included an editor of the *Los Angeles Times* and a Chancellor of the Chicago Law School. Human testicles being in limited supply, the transplanters used chimpanzee, monkey and goat glands. In Europe, the leading figure was a surgeon, Serge Voronoff, who like Metchnikoff was a Frenchman of Russian extraction. Voronoff combined his scientific enthusiasm with a missionary zeal to revitalise the passions of older Frenchmen and make up for the appalling losses of younger blood in the trenches of the First World War. However, in the late 1920s science began to catch up with the gland-grafters. The transplants were shown to have resulted only in scars – if nothing worse – and by the 1930s the major concern about testicle transplants was with lawsuits.

More recently, the torch of cross-species Wonka-Vite grafters has been taken up by Paul Niehans and his followers, with injections of foetal sheep and goat cells. These treatments, under the name of 'cell therapy', are still available at considerable expense from certain private clinics. Personally, there is no way I would ever let anybody inject sheep or goat cells into any part of my anatomy. Not only is there a risk of adverse immunological reaction, or of cross-species transmission of pathogens, but what good can it possibly do? The immune system is designed to recognise and rapidly destroy foreign cells.

One can only suppose that advocates and clients of cell therapy believe that young cells can miraculously rejuvenate the old ones, a little like the rejuvenation of King David in the Old Testament, who was perked up by lying with a young virgin and inhaling her breath, even though we are assured that, in the biblical sense, 'he knew her not'. Remember the classic experiment of Hayflick and Moorhead, encountered in Chapter 7, in which they grew young cells together with old cells in order to demonstrate that cell ageing was intrinsic to the cells, and not due to defects in the culture medium? They found that the old cells stopped growing at their due time regardless of the fact that each old cell lay beside a young companion.

Just because sheep and goat cell injections make no sense, we should not dismiss entirely the idea of using transplanted cells to treat problems of old age. In 1988, headlines around the world heralded a claim that brain tissue from aborted human foetuses had been used by surgeons in Mexico, Sweden and Britain to produce spectacular recovery in a number of cases of Parkinson's disease. If transplanted cells can work in Parkinson's disease, might they work in Alzheimer's and Huntington's disease too? In all three diseases, the root cause of the problem is the death of neurones that the body, left to its own devices, cannot replace.

Putting human neurones into a human body makes more sense than injecting sheep or goat cells, but it is still highly controversial. Not only is there the practical problem of securing the necessary foetal tissue, with all of the important ethical issues that surround such a sensitive source, but there is the question of whether transplanted neurones, even foetal ones, can really do anything useful in a damaged brain. Despite the occasional success story, when a patient has felt significantly better after treatment, the early claims of spectacular recoveries have come to ring a little hollow. There is an acceptance that, for the time being at least, cell grafting is unlikely to help the more advanced cases of Parkinson's disease. Unless there is a major breakthrough, this costly and highly experimental procedure is likely to find limited use.

Further in the future is the tantalising, but even more speculative idea of growing your own cells in tissue culture and using these to carry out cell grafts. The difficulties with this procedure are immense. The cells that need to be replaced in the case of brain diseases are neurones. But neurones do not grow in culture. Therefore the first step would be to identify cells – perhaps a kind of stem cell – that *can* be cultured and that can be induced to become a neurone when we want them to. The second step would be to get the cells to exactly the place where the new cells are required and to engineer the signals to make them become nerve cells when they are there. The third step would be to ensure that they are able to make the right connections with the existing nerve cells. It is conceivable that one day all of these steps may be possible, but the reality is that we are here dealing more with science fantasy than with science fact.

Returning to the present, the huge interest right now is in whether drugs for Alzheimer's disease can make a significant impact on this condition in the next 5–10 years. No one is yet hoping for a cure, the reason being that it is hard, if not impossible, to restore normal function to brain networks that have already been destroyed. But can disease progression at least be slowed? Even better, can disease onset be delayed?

Because Alzheimer's disease strikes most commonly among those who are already old, just a modest delay in the average age of onset would have a huge impact on the number of sufferers. All right, the reason for this is not a joyful one: more of us would die of something else before we had the chance to develop Alzheimer's disease. But for many of us, this is what we might choose.

The first drug to be licensed for the clinical treatment of

Alzheimer's disease was tacrine, approved in 1995 by the US Food and Drug Administration. By then tacrine was a 50-year-old pharmaceutical that had been on the shelf long enough that patent protection had run out, and which had been used experimentally on Alzheimer patients for over a decade. Tacrine is an acetylcholinesterase inhibitor, which means that it acts to inhibit the chemical breakdown of acetylcholine, an important chemical messenger of neurones, which gradually declines in the brains of Alzheimer sufferers as the disease progresses. No one knows if the acetylcholine decline is really a cause of the loss of brain function in Alzheimer's disease, or if it is just a signal of the damage that has occurred, but in the case of a disease as serious as Alzheimer's, it was worth clutching at what might prove to be a therapeutic straw.

Unfortunately, tacrine's performance in clinical trials was not particularly impressive, to the extent that a poll of practitioners and researchers by the *Harvard Medical School Health Letter* revealed little expectation that it would turn out to be a useful treatment for the disease. Tacrine suffers the further drawback that it cannot be tolerated by all potential users and can cause liver damage.

So the field was wide open when a second drug, donepezil hydrochloride, marketed under the brand name Aricept, was approved for treatment of Alzheimer's disease in 1996. Clinical trials of Aricept showed some improvement in the symptoms of mild to moderate cases, restoring some mental abilities for about 6 months, and having fewer side-effects than tacrine.

More recently still, a study published in 1997 examined two antioxidants, vitamin E and a drug called selegiline, and found that either of these could extend the active life of patients with mid-stage Alzheimer's disease for up to 2 years, allowing them to cope unassisted with basic tasks like bathing themselves and getting dressed.

The drug companies, of course, recognise the enormous profits to be reaped with the first really successful drug to treat Alzheimer's disease, but sadly, the breakthrough may be a long while coming.

The difficulty is that, unlike a drug like penicillin, which can wipe out an invading bacterium and leave healthy tissue to recover, a drug for Alzheimer's disease will have to overcome the massive destruction that has occurred already. The biology of the disease just does not offer the prospect of a simple 'cure', even when we have harvested all of the scientific fruits of genetic and other research into why it happens.

A cure will be elusive, but we can at least hope for better prevention. Much excitement has come from the discovery that familiar pain-killers like ibuprofen, drugs that belong to a category known as non-steroidal anti-inflammatories, appear to have an important impact on the risk of disease. This finding first came to light when it appeared that arthritis protected against Alzheimer's disease. What really seems to have happened is that those who were taking ibuprofen for their arthritis were reaping an unexpected side-effect. They were being protected against Alzheimer's disease.

Ibuprofen, like tacrine, vitamin E and selegiline, has been around a long while. It has a drawback, as do other non-steroidal anti-inflammatories, that many people cannot easily stomach it. It can, literally, cause ulcers and bleeding. So the search is now on for new drugs that have the benefit of ibuprofen, but not the side-effects. Once again, it may be a while before these emerge. Any good prophylactic to ward off the evil of Alzheimer's disease would in all likelihood need to be taken long before the disease develops. Think about it. Even when such a drug is made, it will take a good many years before we can really know how well it works.

Wonka-Vite for the brain, then, is likely to be some way in the future, though that is one forecast I would love to have proved wrong. But what about other aspects of ageing?

The monkey gland experiments were, with hindsight, something of a joke, but the idea was not really so daft. What Voronoff and others recognised was that male hormones declined with age and that, if the level could be boosted again, perhaps some of the loss of function in old age could be reversed. Male hormone replacement therapy is beginning to catch on. Recent studies suggest that injections of testosterone once a week, or newer treatment with testosterone skin patches, can help older men stay leaner, stronger, happier and more virile (although a downside may be increased risk of atherosclerosis and prostate cancer). Add to this the effectiveness of Viagra as a treatment for impotence, and some of men's anxieties about getting older can be allayed.

Hormone replacement therapy for women has an even stronger biological rationale, in view of the abrupt shut-down of ovarian hormones that occurs at menopause. Millions of menopausal women now take oestrogen hormone replacement therapy (HRT) to combat hot flushes and the other unwelcome side-effects of the change. There is strong suggestive evidence that HRT brings other benefits in the form of reduced post-menopausal bone loss, reduced risk of breast cancer and of cardiovascular disease, and even a degree of protection against Alzheimer's disease. On the minus side, there is some suggestion that HRT increases the risk of ovarian cancer.

The niggling doubt, however, in most of the HRT studies to date is to know whether the women who elected for a course of treatment were already more active in their pursuit of health and perhaps less at risk of disease than those who did not. Careful as the researchers were to minimise the sources of bias in their studies, data from carefully controlled long-term clinical trials will be needed before the specific effect of HRT itself is known for certain.

Three other hormones have received considerable attention over the past years for their possible effects on the ageing process: growth hormone, dehydroepiandrosterone (DHEA) and melatonin. The rationale for the claims made on behalf of each of these hormones is that production of the hormone declines with age. Boosting the level of the hormone in old age back to something like the level in youth might, it is suggested, ameliorate or reverse some aspects of ageing.

Hormones are chemical messengers by which cells and tissues

within the body signal to each other, sometimes at long range. Because hormones are synthesised inside cells and because, as we saw in Chapter 8, a lot of what goes on inside cells gets corrupted with age by all kinds of damage in the cell's molecular machinery, it is not surprising that hormone production may decline. Hormone signals also need to be interpreted and acted upon by the target tissues. The machinery for responding to hormone signals also deteriorates with age. So there is indeed a case for arguing that adding to the declining hormone levels in old age may, conceivably, do some good.

There is a counter-argument, however. The root cause of the problem – if problem it is – is the decline in what goes on inside cells, by way of sending and receiving the hormonal signals. There is no obvious way that adding more hormone could alter the fundamental processes of ageing, and it could even make matters worse. The real test of the claims for hormone supplementation lies in being able to demonstrate a genuine benefit that results in sustained improvement in old age.

In the case of melatonin, neither the hypothesis nor the data are at all convincing. Melatonin's primary biological effect seems to be on the circadian rhythm, the biological cycle that keeps us working to a 24-hour clock. Melatonin appears to have a specific role in regulating the sleep–wake cycle. The hormone is produced by the pineal gland, which is a tiny structure located deep within the brain. Melatonin is made during the night-time and its production can be set back by bright lights in the evening. Its production reduces with age, and this may be partly why older people experience more insomnia.

Just why melatonin should be a modulator of ageing, as its recent hype has suggested, is not at all clear. There is a world of difference between the biological timing of the daily cycle of life, and the biological timing that is involved in ageing. An alarm clock set to wake you each morning is a 24-hour timing device. The fact that the same alarm clock might after years of good service become a little erratic due to mechanical wear and tear in its mechanism does not imply that the alarm clock also serves to measure this longer timescale. Melatonin is a weak antioxidant, but its contribution to the body's overall antioxidant capacity is negligible. There are no convincing data to suggest that melatonin has anything to do with ageing.

DHEA's role in ageing has also been the subject of some wildly exaggerated claims. It is the antidote to ageing, say some suppliers. The fountain of youth in a bottle, say others. DHEA is a steroid hormone produced by the adrenal glands in both men and women. It plays a role, yet to be fully defined, in the production of other hormones, particularly oestrogen and testosterone. DHEA production falls with age in both sexes, but more dramatically in women. Given that oestrogen and testosterone play important roles in female and male biology, and given that hormone replacement therapy with both these hormones has significant effects in ageing, it will not be at all surprising if DHEA supplementation also has biological effects in old age, some of which may be beneficial. This is a far cry from being the fountain of youth in a bottle.

Growth hormone is, chronologically speaking, the senior partner among the hormone contenders for the Wonka-Vite title. In 1990, a scientific report described how a number of elderly men had gained muscle mass after treatment with the drug. As its name suggests, growth hormone is a potent stimulator of growth in children, and in later life it continues to be produced according to a rather strictly regulated 24-hour cycle, albeit at a declining level. It acts to stimulate cell division and is thought to be important in tissue repair. The problem is that its effect on muscle mass has not been shown convincingly to translate into any real benefit in terms of enhanced strength and endurance in older people. Serious concern has also been expressed about side-effects, including an increased risk of cancer. Growth hormone is strong stuff, and strong stuff should be used with care. Rocket fuel may make an old banger of a car go, but for how long?

The thing that is most worrying about the way that some of the hormone therapies for ageing are being touted and sold is that the claims pay such scant regard not only to the nature of the ageing process itself, but also to what is known about how hormones work. Hormones are potent biological agents, often with complex and multiple effects, and often released within the body on a tightly co-ordinated schedule. Not the sort of thing you would expect to work by popping a pill or two of a nutritional supplement of somewhat uncertain provenance into your mouth and swallowing. Which may, of course, be just as well.

We may smile at the farmer in Navrongo who sprinkled the content of his two antibiotic capsules on to his septic foot, but when it comes to misuse of unproven, often ill-conceived 'remedies' for ageing, it is quite staggering what goes on in the supposedly better-educated parts of the world.

A sector of the Wonka-Vite industry that used to be notorious for its extravagant claims, but which has cleaned up its act a lot in recent years is the cosmetics industry, self-appointed scourge of the wrinkle. What the skin cream makers seem to have realised, aided perhaps by legislators who rule that you should not promise what you cannot deliver, is that combating the *signs* of ageing is good enough. Combating the real thing is rather more difficult.

The other factor that has no doubt engendered a greater honesty in skin cream marketing is that what you get is what you see. There is no point in offering a product for sale that has all sorts of bogus claims made for it, if you can actually see that it does not work. Most customers will not pay a lot of money for it more than once – which might explain why some of them are so expensive!

Anti-wrinkle creams mostly work by using chemistry to affect the appearance of the skin, for the most part by plumping and rehydrating dried-up cells. In spite of the fact that manufacturers have wisely dropped claims to tackle the intrinsic ageing of the skin, some packaging still carries a whole lot of scientific gobbledy-gook designed to impress the purchaser.

A new development is the range of skin creams that use retinoic acid (vitamin A) as an active ingredient. Retinoic acid is a natural substance that has powerful effects on cell differentiation. For example, it affects limb development and regeneration in amphibians like toads and salamanders. A skin cream containing retinoic acid will not tackle the underlying damage that has occurred through intrinsic ageing in the inner workings of the skin cells, but it does appear to have some effect on the photodamage caused by sunlight.

One of the effects of these creams is on the cellular composition of the skin. Because retinoic acid prompts cells to differentiate, and because cancers tend to form from cells that have dedifferentiated – that is, reverted to a more germ-line state – it has even been suggested by some advocates of the new creams that they may cut cancer risks by whipping pre-malignant cells back into line.

A source of Wonka-Vite that many see as the great hope for the future is gene therapy. Each new discovery of a gene that affects the ageing process is reported with a frisson of excitement that, perhaps one day soon, we will unlock the key to extending human life span. Well, maybe. So far the successes of gene therapy in establishing lasting therapeutic benefit have been few, even in cases that are most obviously amenable to this approach.

Current research on gene therapy is mostly directed at the correction of inherited genetic disorders such as cystic fibrosis. The idea is to introduce the normal gene into enough cells in the affected individual that their tissues can function at least part-way normally. The commonest technique that has been tried is to package the gene into the genome of a harmless virus that serves as the carrier, or vector, and then get the virus to infect the target cells. If the virus succeeds in infecting the target cells in sufficient numbers, and if the gene works in these cells, then the hope is that it will produce the protein that is defective or missing. The limited success to date has revealed pitfalls in each of these steps. Even if the virus does infect the target cells, and even if the gene works as it should, how long will the effect last? If the treated cells get deleted as part of the normal tissue turnover, you are back to square one and have to do it all again. There is also the chance that the virus vector will itself cause mutation in a target cell.

When gene therapy for inherited disorders has been made to work, and let us hope for success just as soon as possible, we will still be a very long way from producing genetic Wonka-Vite. The problems confronting gene therapy against ageing are formidable. First, we will need to know exactly what we want to treat in our gene therapy of the future. We will need to discover a great deal more than we know now about the gene networks that affect the ageing process. We will also need to carry out some extensive studies of how changing individual genes in these networks will affect the process as a whole. We may find that there will be side-effects that we want even less than our current ageing process.

Secondly, we will need to identify exactly how we will effect the gene alteration, whether we will use germ-line or somatic gene therapy, and what gene delivery system we might use. Thirdly, if and when we ever get this far, we will have to wait 100 years or so to see if it works! The bottom line is that there is no guarantee now that we will *ever* be able to engineer genetic Wonka-Vite.

Gene therapy to combat some of the diseases of ageing is a more feasible prospect, but even here we really do not know yet the prospects for success. When genes are discovered that predispose us towards an age-related disease like Alzheimer's, we welcome the advance for the insight it brings into how the disease is caused, not because we have any serious expectation that this is a gene we might think to alter.

The prospect of following Grandma Georgina even part of the way on her journey towards Minusland seems a long way off, but actually there is one rather surprising kind of Wonka-Vite that we are coming to understand a little, and this is Vita-Wonk itself. The idea that that which harms you does you good is a strange one, but in a way it makes sense and it has a long scientific history. The idea even has a name: hormesis.

If you subject a system to stress, and if the system has the capacity to respond to that stress by switching on, or turning up, its protection mechanisms, then you might end up better protected than you were to start with. Traditional healers in West Africa used to use a similar trick to impress their clientele. By regularly injecting themselves with a little bit of snake venom, they could build up resistance to snake bite. Nothing could be more effective in suggesting magic powers than to take a full-fanged bite with impunity.

Some years ago, biologist Joan Smith-Sonneborn at the University of Wyoming showed that shining ultraviolet light on little unicellular creatures called *Paramecium* could make them live longer. This was surprising because what the ultraviolet light was doing was damaging the DNA inside the *Paramecium*. *Paramecium* divide only a certain number of times before they die out, rather like human fibroblasts and the Hayflick Limit.

The way *Paramecium* avoid going extinct is by having sex. There is good reason to believe, as we saw in Chapter 13, that one of the important functions of sex is to help keep the DNA in good shape. What the ultraviolet light did was to allow the *Paramecium* to go longer without having sex and without dying out. The way it did this, Smith-Sonneborn found, was by stimulating the *Paramecium* to turn up their DNA repair mechanisms.

In much the same way, my colleague Gordon Lithgow at the University of Manchester has shown that you can take nematode worms and stress them by gently turning up the heat. In this way, you can build up their capacity to deal with all of the junk proteins and other cellular damage that the heat produces. If you get it right, as Lithgow has shown, you can prime them to live 40 per cent longer. This is not something to be tried at home!

In case I have disheartened you by suggesting that the search for Wonka-Vite is a hopeless cause, let me close this chapter by saying that nothing is further from the truth. We *are* living longer and longer, and there is encouraging recent evidence in the United States that the period of disability before death is getting shorter. In other words, the health span may be lengthening even more than the life span. But you would be kidding yourself to think that long life can be bought in a bottle.

CHAPTER 16

Making more time

Jenny kiss'd me when we met,
Jumping from the chair she sat in;
Time, you thief who love to get
Sweets into your list, put that in!
Say I'm weary, say I'm sad,
Say that health and wealth have missed me,
Say I'm growing old, but add,
Jenny kiss'd me.

Leigh Hunt

If you have gathered nothing in your youth,
how can you find anything in your old age?

The Jerusalem Bible, Ecclesiasticus 25:3

We closed Chapter 1 with the assertion that knowledge of the science of human ageing can help us all to approach old age with a deeper understanding of what is happening to us, and can, to some extent, enable us to exercise a greater degree of personal control. It is time to take stock of that knowledge and to ask *how* it can do this.

You only get one soma in this life, so if you want it to last, you had better take good care of it. The fact that our life span is *not* programmed into us from the time of our conception, and can be influenced by how we choose to live our lives, brings power and responsibility in equal measure. Our genes treat the soma as disposable. So far, we lack the means to outwit them. However, we do have the means to make the very most of our biological potential.

I have always felt that the motor car industry would privately prefer that your car lasts 10 or 12 years at most. They want you to buy a new model and to keep them in business. There is much fun to be had in buying a new car, but if you are attached to your old one, you can go a long way towards confounding the wishes of the motor car manufacturers by regular maintenance, careful driving and frequent wax polishing.

You cannot, of course, prevent serious accidents from causes outside your control, any more than you can prevent fatal diseases, like cancer, that arise spontaneously through the action of chance. As soldiers in the trenches of the First World War wryly observed, it is not the bullet with your name on it that is necessarily the biggest worry, it is the one labelled 'To whom it may concern'.

Personally, I would like to live for as long as possible. I make this statement unhesitatingly now, being in a state of good health and happiness. This is not to say that my life is without its burdens and cares. But the good things sufficiently outnumber the bad that *for the time being* I am quite clear about my preference. This is something that in all our lives can change.

During the years of alcohol prohibition in America, the late W.C. Fields, comedian and renowned alcoholic, used to joke that he would rather die than live in Philadelphia, bastion of the temperance movement. For his final joke – though perhaps it was no joke at all – he chose for his own epitaph the reflection: 'On the whole, I'd rather be in Philadelphia'.

As ageing progresses, life becomes increasingly more unpredictable. Ill health, frailty, bereavement, poverty, depression and loneliness are common hazards. Each of these can levy a heavy toll on the quality of our lives. Some cope best by denying that these things lurk in the future, but for most of us it does no harm, and actually may help us a great deal, to think ahead a little.

People who in youth had frequent contact with older people, especially those who had cause or opportunity to be care-givers, seem to find it easier to be at the receiving end of care when they themselves grow old. Sociologists Toni Antonucci and James Jackson at the University of Michigan have likened this process to a 'social support bank' into which you make payments when you are young, and from which you withdraw care when you get old. Most of us do not like to feel that we are on the receiving end of support from family and close friends without being able to give something in return. We strive for reciprocity. But as we become more frail, it becomes harder to reciprocate. This is where the idea of the social support bank comes in. If we can reconcile our current inability to provide support with the feeling that, over our life-course, the exchange has worked out more-or-less equally, we can more easily set our minds at rest. The more one has invested in the bank, the less one feels that one is a burden, undeserving of support, when infirmity strikes. This issue of justice between the generations is going to be one of the biggest problems that wealthy societies around the world will need to resolve in the coming decades, as the fiscal implications of the demographic revolution bite deeper and deeper.

Even in the traditional gerontocracies, such as exist in and around Navrongo, the relations between old and young have never been quite as straightforward as they seem. The 'old man' who granted permission for my photograph of the painted house could not have asserted his great authority by himself; he was much too frail. When I asked young Ghanaians what they really felt about their traditional tribal structures, I found that the genuine respect for their elders was much tempered with negative feelings. One wrote the following candid account:

> The old have a significant role to play in the life of the society. They represent life, history, experience, the importance of the past in the present. They can be rocks of assurance amidst the quicksands of rapid change.
>
> But in our present society they tend not to get their well desired attentions due to the constant drift of the youth and educated from

the poor rural areas to the rich urban centres, the rapid eradication of illiteracy, money economy and business techniques – which have left many old confused, disgusted or else afraid.

It is the old in our society who are the most experienced medicine-men/women, herbalists, soothsayers, spirit owners, diviners, controllers of the African-black-power, etc. Though they are highly respected for these important roles they play, people turn to fear them and are very careful towards them – for fear that they may do whatever evil to them. They therefore feel unsafe in the society because they are mostly avoided.

For the feeling that our old are just about going to the grave people/relatives do not give the best things in life to them. The old also have a careless attitude to themselves. Mostly they wear the most wretched clothings – they look dirty for most of the time.

Though so many bad things are associated with the old in our society, they are always protected for they are the leading patrons in every traditional activity.

How long can a system founded on such ambivalence last against the rising pressure to transfer initiative and power to the young? And at what social cost will these changes come?

In developed countries, older citizens have comprehensively lost ground to youth power in many areas of life. Migration has fragmented traditional communities, as the young move elsewhere for work and new families move in. The rapid pace of technological change has altered life, so that many older people feel unsure of the relevance of their store of experience and wisdom. Grandchildren teach grandparents how to work gadgets, when once it was the other way around. Social attitudes are learned as much from television as from the extended family. Shopping patterns have changed in favour of supermarkets and malls, and against the smaller neighbourhood stores, all of which tends to disadvantage those whose mobility and physical strength are limited by age.

Security for many is afforded by the simple fact that during their long lives they have managed to accumulate enough capital to ensure financial self-sufficiency in retirement. However, increasing numbers of older people lack this buffer and must rely on what the state feels it can spare them.

Older people worldwide must lobby vociferously for their own cause, and those who hope to live long enough to become old must support them. As a society, we must develop greater awareness of these issues and develop the moral courage to face them openly and honestly. Health services will come under increasing pressure on resources and if we are not to be ageist in our allocation of treatment options, we need to ask some searching questions of ourselves, like those asked of the helicopter pilot in Chapter 2. Rescuing holiday-makers from a sinking yacht is but another way of looking at how we might draw up priority lists for kidney transplants, for example. Major surgery is riskier at older ages, but not as much riskier as all that, if the older person is in good health. Our bodies have a considerable capacity for effective healing and repair, even in extreme old age. When old people are asked their own preferences over whether or not to choose aggressive treatment options for life-threatening or disabling conditions – such as surgery or chemotherapy – versus merely making things as comfortable as possible, they are almost as likely as younger people to opt for the more drastic treatment in the hope of a cure. Their choice may be tempered, in a situation where resources are limited, by concern not to deprive a younger person of the chance of treatment; but they do not want to be denied the option *merely because they are old.*

Ageism should make us angry, like racism and sexism. In England not so long ago, a tragic accident happened where an elderly woman suffered a heart attack and died while driving her car. The car struck and killed a young mother and her daughter. 'What was she doing driving at her age?' demanded the national press headlines.

Well, why not? The elderly woman was a careful driver with an excellent record of safety. As far as she knew, she was in good health and had been certified fit to drive. Vastly greater numbers of fatal accidents are caused by young men and women *knowingly* driving too fast or while drunk, by business men and women *knowingly* driving too far and falling asleep at the wheel, by drivers of all ages being distracted through *choosing* to use a mobile phone, and by *culpable* neglect of vehicular maintenance. And does not our car-centred culture make life rather difficult for older people if they do not drive?

These wider social issues extend well beyond the scope of this book. Nevertheless, understanding the science of human ageing can help us to prepare more effectively for old age. If we know what ageing is and what causes it, we can avoid it creeping up on us unawares, and we can fight the process each step of the way. It is never too late to start, but it is worth remembering that some things are best done sooner rather than later.

Diet is a good place to begin. Eat what you enjoy – to do otherwise might make you live longer, but for what purpose? – but train yourself to eat fewer calories, if you can. I find this hard because I like food enormously. But there are all kinds of tricks you can play on yourself, some as basic as using smaller utensils so that a reduced portion does not look quite as lonely. Set realistic targets, and try as far as possible to replace Vita-Wonk foodstuffs (sugar, fat, etc.) with Wonka-Vite foodstuffs (fruit, vegetables, etc.). Top up your vitamins (particularly C and E) and trace elements with nutritional supplements if you feel you need to, but aim to get as many of these from natural foodstuffs as you can. Fish, particularly oily fish, and red wine should be on the shopping list.

These measures are best adopted as early in adulthood as possible, but it is never too late to start. However, dietary change should not be imposed heedlessly on the very old. Some years ago, I had the enormous pleasure of travelling to a conference of ageing and nutrition and meeting Elsie Widdowson, one of the leading nutritionists of this century, who was then in her nineties. Widdowson spoke strongly against trying to impose a change of eating habits at an advanced age, and there is a good deal of force to the argument that, if you have lived a long time already, your diet

cannot be far wrong. I travelled back to England with her and
during the flight she entertained me with tales of the Institute
where I then worked and where she had been employed before I was
born. At this Institute in my own time there worked a senior
scientist, still doing excellent and highly regarded research,
although some decades past retiring age. This individual ate exactly
the same modest lunch – one slice of wholemeal bread and a small
bowl of dessert – every day.

After nutrition, exercise is the obvious place to continue.
Exercise is famously good for your muscles and cardiovascular
system, as well as for controlling weight. Walking is fine exercise if
you do not have the amenities or inclination for other sports. Using
stairs instead of lifts is a good idea, not only because the exercise is
a bit more taxing than walking on the flat, but also because the
slight jolting that you get as you walk downstairs stimulates your
osteoblasts to build more bone. Battling the force of gravity is
somehow important for bone maintenance, as astronauts who
spend long periods in the weightless conditions of life in space find
to their cost.

Many old people like to move from houses with stairs into
bungalows or apartments as they become more frail. This can ease
life considerably and cut the risk of falls, but it is worth
remembering that the loss of the exercise gained from going up and
down stairs needs to be compensated for in other ways. Research
has shown the enormous health benefits of a programme of regular
exercise for older people, and it is never too late to start. Because
the power of muscles declines with age, many elderly people
(especially women) are required to operate at or near to the limit of
their physical powers, even when carrying out everyday activities
like rising from a chair or carrying a basket of shopping. In this
sense, they are not so very different from the Olympic athlete who
must train regularly to keep in shape. There are plenty of effective
exercises for older people that can be done even in a seated position.
Dawn Skelton, a British researcher into the benefits of exercise for
those aged 75 and older, has shown that such training can result in

an improvement in strength equivalent to 15–20 years' 'rejuvenation'.

Exercise is good not only for its general effect on muscle strength and the fitness of the heart and lungs, but also for combating ageing itself. There is new evidence from the University of Newcastle upon Tyne in England, which suggests that vigorous exercise can actually slow the intrinsic processes of cellular change that cause ageing in muscles. Veteran athletes who continue running into old age accumulate fewer mutant mitochondria in their muscle fibres than the rest of us, perhaps because they push their mitochondria (the energy-producing units within cells) hard, and because this forces a strong selection to keep the populations of mitochondria within the muscle cells in good shape. Otherwise, the mutant mitochondria can just accumulate with no selection force to get rid of them.

It is worth taking up or continuing regular exercise if you can. I have run just two marathons in my time and during the first of these I ran most of the way alongside a man whose 70-year-old legs matched my 30-year-old legs stride for stride. But it is important to avoid muscle and joint injuries by warming up and performing stretching exercises before and after major exertion. I used blithely to ignore this advice when I was young, even though I ran competitively. Once I reached my forties, however, I quickly found that running without warm-up was a rash thing to do.

Another form of exercise (and more!) is sex. It is one of the greatest crimes against the aged that enjoyment of sex has been hijacked by the younger generations, who sometimes seem to regard sex and sexuality as their exclusive preserve. Sex is for everyone, if they want it, and there have been many studies that suggest that to continue a healthy sex life into old age can produce significant physical and psychological benefits.

There are, of course, physical impairments in the functional aspects of sex that become commoner with age, but most of these can be overcome or coped with, just as so many other aspects of the ageing body have to be overcome or coped with. The 'use it or lose it' principle applies to sex as much as to other aspects of the ageing process.

One of the factors inhibiting sexuality with ageing is undoubtedly the brain-washing that all of us experience which says that the older person is less sexually attractive. This needs challenging. Another myth that needs debunking is the idea that sex at older ages is unsafe, because of physical frailty. Well, really! If there is a benefit that has come from the grim experience of the AIDS epidemic, it is the idea that safe sex is OK. If – and in my view it is a big if – the practice of certain forms of sex becomes unsafe because of physical frailty, then let us use our imaginations.

Exercise for the mind is as important as exercise for the body. There is some evidence, as we saw in Chapter 9, that regular mental exercise, of the type that chess-players and crossword puzzle-solvers enjoy, can slow some aspects of memory loss. There is also evidence that advanced education is associated with slower onset of Alzheimer's disease. One of the best ways to exercise the mind is to keep learning. The University of the Third Age is an international organisation specifically directed at lifelong learning past normal retirement, and there are many other opportunities to acquire new skills and interests throughout life.

A nice example of someone combining exercise and learning was a man I met recently who was strapping himself into his hang-glider harness at age 72. I must have looked a little surprised because he explained that this was something he had always wanted to do, but had not done before now because of family responsibilities. His children were fully independent and his wife had recently died, so he had just taken a course in flying a hang-glider and was about to go solo.

There will, in time, come to be more, perhaps much more, that we can do to stave off the ravages of ageing. It is likely that we will see mortality rates in old age continue to fall, and the health span go up. Since 1950, the numbers of centenarians in many of the developed nations have more than doubled each decade, according to a database kept by Finnish demographer Vaino Kannisto and

British demographer Roger Thatcher at the University of Odense in Denmark. The mortality rates in the oldest age groups are continuing to fall as old people enjoy better living conditions and reach old age in better shape.

The measures listed above are available now, but how many of us will take them to heart and get the best from our somas? Some do already, and perhaps greater numbers will follow each year. As we learn more about the genetic factors that predispose us to health problems in later life, we will be able to tailor our actions more effectively.

For many people, however, the effort and personal sacrifice involved in diet and exercise are just too much to invest for the uncertain gains of future health in old age. We tend to have a short time horizon and to discount future benefits accordingly. *Carpe diem*!

So the great hope of many is a simple cure for ageing, the quick fix, the elixir or fountain of youth. Well, it may not exist. And if it does, it may be a very long time coming.

The rate of progress in gerontology is exciting and all kinds of benefits will come from this research, in terms of both fundamental knowledge and practical application. To put this in perspective, we might compare it with other fields of medical research.

AIDS is an illuminating example. We know nearly everything there is to know about HIV as a virus. We know the entire sequence of its genome. We know all of its proteins and have a pretty good idea what they do. We have a huge amount of information about the action of the virus in infected individuals. But whether and when we might be able reliably to cure this single infectious disease, or even to vaccinate against it, is still very uncertain.

Cancer is another example. When United States President Nixon declared his 'War Against Cancer' in the early 1970s, there was every expectation that by throwing enough money and effort at the problem, the scourge of cancer could be banished. Well, what happened? Billions of dollars of research money have poured into cancer research worldwide, and millions of person days of research effort have pursued this elusive prize. There has been real progress and many cancers are significantly more treatable than ever they were before. But once again, whether and when we might eradicate cancer is uncertain. Simply declaring war on a problem does not guarantee a solution and creates a false expectation of success.

Ageing is a bigger challenge than AIDS or cancer, and ageing research is still in its infancy. From understanding ageing we will better understand many age-related diseases, like cancer, Alzheimer's disease and osteoporosis. At least we now have a good idea what to look for, and every year we have better and better tools to search with. Progress is being made faster than many believed possible just a few years ago. Undoubtedly the future benefits from this research will be huge, but as yet the details are unknown.

Scientists are accustomed to exploring the unknown, and they do so on behalf of the society of which they are a part. Society shares in the rewards and the risks. It serves no useful purpose to pretend that the deep secrets of ageing will come easily. But the more we learn, the more reliably we will be able to anticipate future discoveries.

In the meantime, there is much that we can do to make more time for ourselves. We may not yet be able to add many more years to our lives, but we can certainly add a lot more life to our years. We can begin by taking a good look at the ways that we use the time available to us. The average American adult, according to the Television Bureau of Advertising, spends 28 hours a week watching television (some 25 per cent of waking hours). In spite of the excellence of some programmes, it is hard to believe that all of this time in front of the television is well spent. A woman in Navrongo may be occupied this long each week pounding millet and fetching water from the borehole. A man may spend it walking to and from work. It is a sad thought if all of the labour-saving devices of the developed world have freed up this time for so little purpose.

Free time was a luxury in Victorian times too. So precious were the free hours in the working weeks of the workers in mine, mill and factory that for many they were carefully invested. They were invested not least in hard-won self-improvement in the Mechanics Institutes, the brass band clubs and the like. Nowadays we have freedom to use and manage our time as never before. But freedom, much as we want it when we are denied it, is a daunting responsibility when we have it.

Freedom and time are relative concepts. They depend upon your point of view. A day in the life of a child is a different experience to a day in the life of an adult, just as no doubt a day in the life of a mouse is different from a day in the life of an elephant. In a very material way, the rate at which time *appears* to pass must to some degree be influenced by the rate at which new experiences are encountered relative to the store of experiences held in memory.

The span of time between a child's fifth and sixth birthdays is enormous when set against that child's experience of life already, whereas the span of time between a person's seventieth birthday and their seventy-first birthday is, relatively speaking, far less. The disparity can be much reduced, however, if the 70-year-old sets new targets for learning and fresh experience.

One new experience that is ahead of each of us, and that looms ever nearer as we grow old, is the actual experience of dying. We might do well to keep alive our openness for new experience in order to cope better with this challenge when it comes. In earlier times, death was an ever-familiar presence. Considerable importance was attached to the idea of leaving this life as one has lived it, i.e. well. It is very striking that children with incurable diseases approach dying with a courage and a dignity that can teach us a lot. Recently, a documentary was shown on our television screens that contrasted a Texan millionaire's quest for immortality with a child dying from Hutchinson-Gilford progeria. Progeria is a rare genetic condition of accelerated ageing that runs an even faster course than the Werner syndrome that we encountered in Chapter 7, though its cause is still unknown. The child displayed a mature acceptance of the inevitability of her early death, talking regretfully but openly about what was happening to her. Her touching grasp of this reality contrasted poignantly with the fantasy that appeared to have the millionaire in its grip.

Freedom makes us individually responsible for our choices and our actions. Is this why we so readily drug ourselves into inactivity with low-demand time-fillers when we could do so much? Let us be truly alive, so that when old age finally robs us of our vitality, we may feel that the time of our lives was well spent.

Epilogue

A Short Story

After five days the Capsule had done its work, and Miranda lay dying.

It had been a shock when it had detonated, silently of course, and the world had turned grey. One moment the air taxi, skimming over the northern forests, had been bathed in the rich glow of autumn colours, red, brown and gold, reflecting the low, piercing rays of the newly risen sun. The next moment, or so it had seemed, the sun's gold turned a brilliant white and the forest beneath became a sombre blend of black and greys.

The loss of her colour vision must in reality have taken some minutes to develop, Miranda realised later, but she had been absorbed in her thoughts of the meeting to come and had paid only glancing attention to the glorious view from the window.

The taxi docked at the Quebec terminal and Miranda, thanking the now monochrome driver, had walked the short distance to the little park by the river, the place where she first met Gregor so many years ago, to rethink her plans.

She would of course still attend the meeting of the Council of Timed Ones, which was due to start in half an hour. Her report, incomplete and enigmatic as it was, would not take long to deliver. The Council would find the content unsettling, but as yet there was too little of substance to warrant lengthy discussion. A further study would be authorised, and Miranda could ensure that Juno, her capable deputy, was charged to continue the work of their team. Miranda would be free to leave the Council meeting by early afternoon at the latest, which would allow sufficient time for the air taxi to have her home by nightfall. Inevitably this arrangement meant sacrificing one of her five remaining days to duty, but that would still leave Miranda time to make the necessary arrangements. In many ways, she was glad that the Capsule's detonation, so long awaited, had come while she was busy.

Miranda gazed out over the swirling eddies of the wide river. There was, in truth, little to arrange. Her financial affairs would be settled after her death according to the simple plan she had drawn up long ago. Never one to leave things to the last minute, Miranda had already carefully selected those whom she would invite to attend her dying. The list was a short one. Choosing fitting mementoes for each of them from her small stock of cherished posessions would need careful thought, but even that would not take long. And then, of course, there was Lara.

Lara was the real reason for Miranda's approaching death. And Gregor, too.

'Would Gregor have come?' Miranda thought, as she watched the silvery light sparkle on a stretch of distant water. 'Could he have borne it?'

Gregor had entered Miranda's life a short time after her ninth fraitch, which, she reflected, would put her in her late 220s. Gregor himself was then nearing his third fraitch, which made him about 150 years her junior. Not that it mattered.

Miranda's love for Gregor had taken her by surprise. It had been immediate and deep, eclipsing the previous loves of her long life. Make no mistake, Miranda's earlier loves had lacked neither warmth nor joy. One of them even resulted in the birth of her cherished son, Nico, now one of her closest friends. But the problem, if problem it had been, was that Miranda had always previously held something important in reserve.

Holding back from full commitment was a habit conditioned by the boundless possibilities of an unlimited future. The only known strategy to cope with this awesome prospect, short of mind-numbing drugs and escapist diversions, was to cultivate and preserve an exaggerated love of oneself. In the early centuries after fraitch technology was developed, the emotional burden of long life was poorly understood and the suicide rate grew alarmingly high. Psycho-fraitching of the mind quickly became as important as the regeneration of the cells and tissues of the body.

After their first meeting by the river, and during the heady weeks that followed, Miranda had been startled to discover that Gregor loved her with an intensity and passion that went way beyond all of her previous experience. Not short of passion herself, Miranda found her reserve and self-absorption melting away. She delighted in Gregor's presence and he in hers. When Miranda gave up her farmland home to live permanently in the limestone caves where Gregor had carved his beautiful dwelling, her friends were jolted with the shock. With the quaint exception of the Snuggees, a near-invisible sect that inhabited the far north-east and practised, so it was said, the bizarre habit of 'family living', most individuals preferred to live alone, meeting by choice to share bounded periods of time.

Fifteen full and happy years passed quickly in Miranda's and Gregor's lives, their time occupied with creative work and play. During these years, Miranda's love for Gregor had grown ever stronger and deeper, until the day finally came that Miranda made the decision that would alter their lives for ever. Miranda decided that she wanted to share with Gregor the making of a child.

In a world freed from the necessity of ageing, the making of children had very great significance. Children were still needed to replace those who died from accidents or suicides, but the accidental death rate was so small that their production had to be strictly controlled. The method was simple and stark. Each individual at birth was genetically screened and assigned the right to share in the making of a certain number of children. The usual number was two, but sometimes a smaller number was awarded to limit the spread of harmful genotypes. Exceptionally, a person might be allowed three children if, for example, the recent toll of accidents had been unusually great. The bonus of a third child was awarded by random selection.

To guard against abuse of the quota system and to protect against possible genetic damage to the reproductive cells, which might have a very long wait before use, all fertilisations were carried out in vitro from stored germ cells. Once sufficient germ cells had been removed to cold storage, the gonads were rendered sterile.

To share in the making of a child, a couple would declare their request in a civil ceremony of great solemnity, and following rigorous checks on quota status and genetic compatibility, the fertilisation would be performed. The resulting embryo would then be raised to term either within the womb of the mother or, as was increasingly the custom, in foetal incubators.

For a person with a quota of two, like Miranda, the making of a first child was without major consequence. The parents might choose to participate closely in the rearing of the infant, or they might spend only occasional time with their child, as they preferred. They might do so jointly or, more usually, as individuals. A greater preoccupation with self had weakened the traditional bonding of parents with each other and with their child. In the interests of all, it had become both custom and law that the primary responsibility for the welfare and education of the child rested with the community of which the child would, in due course, become a long-term participant.

However, the making of the final child of a person's quota was an entirely different matter. The birth of this last child signalled the parent's forfeiture of the right to any further fraitches beyond an immediate and final one, at the completion of which a Capsule was implanted. This terminal fraitch delivered the same rejuvenatory effects of the earlier fraitches, but the implanted Capsule imposed a delayed sentence of death. At a random point in time, between 40 and 50 years from the date of implantation, the Capsule would detonate, causing the release of a sequence of neurotoxins that would bring painless death in 5 days. Any attempt at surgical removal of the Capsule would trigger immediate detonation. The bearer of such a Capsule became a Timed One.

It was this fate that Miranda elected for herself when she decided to make a child with Gregor.

Whether Gregor truly understood the nature of Miranda's choice at that time, Miranda never knew. A proposal to share in the making of a child was the deepest token that could be offered to a partner. By custom, each partner's motives for making and either accepting or declining such proposals were respected as the individual and private concern only of the other. Good manners forbade questioning. Rigid convention held that a person's reproductive quota was their own personal secret. Perhaps Gregor persuaded himself that Miranda was lucky enough to have a quota of three, for he knew Nico, her son, well.

But after Lara, their daughter, was born and Miranda adopted the white sash of a Timed One, the light had seemed to leave Gregor's life. Then, a week before Lara's second birthday, had been the accident. Gregor, always more of a risk-taker than most, had gone flying with his sailwing over the Middle Mountains and not returned. By the time his broken body was discovered in a rock-strewn gully 2 days later, his organs were beyond the repair of even the most skilled of the regenerators.

And so had begun the final phase of Miranda's life, in which, as a Timed One, she had access to the Great Archive and began the work she had grown to love.

During her first months in the Archive, Miranda had immersed herself thoroughly in the history of the Dark Years. She read how the scientists of the late twentieth century had at last begun to understand the ageing of the body and had groped their way with crude tools and glimmerings of insight into the deep mysteries of development and regeneration. She marvelled at the progress that had been made, in spite of primitive techniques, during the middle decades of the twenty-first century in the postponement of some of the worst sufferings of old age. She read with incredulity of the war between the maximalists, who valued life extension above all else, and the utilitists, who wanted to impose a crude yardstick by which quality and productivity of life could be measured, with euthanasia applied to those whose tally was negative. She learnt of the brutal triumph of the utilitists in certain eastern regions, where the structures of society had finally broken down under the weight of the demographic revolution, reinforced by the consequences of harsh fertility control. Too few young people could not keep the traditional systems going, and in truth felt little obligation to do so. She read disbelievingly of the devastating competition among private corporations, in which each sought to exploit scientific discoveries for amassing personal wealth. And she read despairingly of the Cataclysm that befell the world when the strains became too much and, in the early years of the twenty-second century, economic and social order broke down altogether.

Ironically, it had been the quest for eternal youth that caused the deaths of so many. Life spans had continued to inch upwards as the diseases of old age were brought increasingly under control. These diseases were not eliminated, of course, but their onset was postponed, and the numbers of those celebrating their hundredth birthdays had grown and grown. But ageing itself had not been affected, and still no one had lived past 130 years.

Then came word of a breakthrough. A remarkable success by a small and secretive research laboratory, Timespan Inc., had engineered mice that, it was claimed, lived twice the normal life span. No special procedures, like near-starvation diets, were needed to produce these long lives. The key lay in the controlled renewal and regeneration of tissues from cryopreserved cells.

Application to humans remained a distant prospect, however. Even the research that might make this faint prospect a reality was deemed morally unacceptable by the World Council on Ethics. Undaunted, Timespan Inc. built a second research facility on a remote stretch of the Caribbean coast of central America under the guise of an exclusive health colony, and began the clandestine recruitment of human clients.

In spite of the intense secrecy with which Timespan shrouded its operations, it became known that a small but growing number of the world's richest people had withdrawn, apparently permanently, from society. Rumour spread that they had joined an immortal élite at the Timespan colony. An angry groundswell of public opinion sought to force an inspection of the colony by the World Council on Ethics. While the Council members debated the issue at length, a direct invasion of the colony was organised by VuNews, the largest of the news corporations, citing overriding public interest as just cause. The results were horrific. Rejuvenation had worked in part, but the control of cell proliferation was imperfect, and gross deformities had resulted. Worse still, rejuvenation of brain function had been a total failure. The mutilated and often demented clients remained voluntary prisoners in the Timespan colony, awaiting the promised refinements that might relieve their sufferings.

The refinements never came. Instead, the universal horror and revulsion produced by the VuNews broadcasts created an atmosphere in which a long-smouldering resentment of scientific interference, as many deemed it, with the natural world, suddenly caught flame. Fanned by the rhetoric of charismatic leaders and armed by a motley assortment of anti-establishment militias, a violent outbreak of hostility was directed at all that could be identified as the product of science and technology. A three-day orgy of rioting and mass destruction disrupted power systems, transport networks and communications equipment across the world.

The records of what happened next were scant, Miranda found. The material damage cannot, in absolute terms, have been very great, but it was more extensive than had ever been envisaged. Few had appreciated until then the knife edge along which the civilised world had advanced, as a remorseless drive for greater and greater technological efficiency had cut away the capacity to recover from such a comprehensive breakdown. Before order could be restored, a disastrous destabilisation occurred. Panicking citizens stripped city stores of their food stocks, and the supplies of fuel for vehicles were drained dry. With no prospect of further supplies being delivered, and with looting and violence spreading, an exodus began that the police and armed forces sought unsuccessfully to control. Those not killed in skirmishes with refugees and looters soon realised the futility of trying to stem the tide and joined the great escape. But escape to what? Rural districts set about defending themselves, their crops and their livestock, but soon they too were over-whelmed. Machine-intensive farming ground to a halt. It was later estimated, Miranda read, that one-tenth of the world's 6 billion inhabitants died in that first desperate month. Three-quarters of the rest did not survive another year, falling victim to cold, hunger, drought or raging epidemics of disease.

Gradually, the remnants of the human race rebuilt a life of sorts. But it was a grim time indeed. Even in the more lately developed regions of Africa and Asia, traditional farming skills had been abandoned two generations or more ago. The terrible mistakes of history repeated themselves. Potatoes, quick and easy to grow, sustained many a community until a new blight revealed again the perilous dangers of monoculture. Millions died in the American potato famines of the 2130s.

Fortunately, a few hard-copy libraries had been preserved through the electronic age, more as historical curios than as practical sources of information. Of these, the collection at the National Library of Medicine near the ruined city of Washington proved the most valuable. Desperate to restore genetic diversity and to breed crops and animals better suited for traditional agriculture, the farmers of the new era immersed themselves in the study of all aspects of biology. Machines were discouraged although not outlawed, provided that lack of critical dependence on mechanisa-tion could be demonstrated. Most important, a thorough working knowledge in at least one branch of the sciences was deemed an essential qualification for citizenship of any of the newly emerging societies.

The harnessing of so much intellectual power to the pursuit of science led to important advances, and the sound application of new discoveries was greatly aided by the deepened public under-standing of the issues at stake. Ethical assessment was an integral component of each programme of research and development, and the involvement of the wider community was easily secured.

Miranda had been almost amused to read in the archives the accounts of the clumsy ethical debates that had occurred, particularly in the latter half of the twentieth century, when non-scientists who eschewed the need to learn science had struggled to understand what the scientists were really aiming to do, and the scientists, so immured in their own narrow researches, were almost equally unappreciative of the worries and misapprehensions that troubled their fellow citizens. Of course, it had not really been funny, Miranda acknowledged. The seeds of the Cataclysm had lain in such divisions, waiting only upon the conditions for their eventual germination.

In the area of medicine, the greatest advances came first in the new cures of cancers, founded as they were in a precise understanding of what it was that made the cancer cells divide and spread. Different cancers needed different cures, but at last they were all brought under control. From the control of cell proliferation came the ability to induce the repair of damaged or amputated limbs, through precise manipulation of regenerative stem cells. The hardest challenge had been the repair of the spinal cord and brain, but in the end even this had been possible.

It was around this time, Miranda had already known, that the first research team had begun seriously to work on a prototype of the fraitch. As any high school student knew, fraitch technology was based on stem cells. More complicated was the way that the stem cells were induced to migrate through the body to find their 'shadows', the pre-existing stem cells within the target tissues. The infusion of the molecular signal that would cause the new stem cells to activate the self-destruct signals on the surfaces of their shadows, and then to engulf and digest their dying remains, was the real moment at which the fraitch took place. No matter how many times Miranda had struggled through the relevant chapter in her textbooks, she knew she still did not really grasp how the renewal of the nerve cells of the brain took place. It was similar to shadowing, but the much harder part was how the new nerve cells were guided to make the same connections with other cells, thus preserving the memories and identity of the person being fraitched.

'Too late to fix that now,' Miranda thought, but she knew that the nerve cell renewal remained the least secure aspect of the fraitch. It was estimated that a good fraitch restored at least 99 per cent of the earlier brain cell connections. The 1 per cent that failed to get reconnected right made for some slight discontinuity, which Miranda knew from experience could be a little unsettling. On the other hand, there were those who liked this element of novelty in a fraitch. By convention, those nearest and most important to you would drop anything if invited to be with you in the days after a fraitch in order to help restore lost memories. They do the same for a dying, of course, thought Miranda wryly.

The early fraitches had been done with stem cells taken from the body, screened in the test tube, and cryopreserved. Now, of course, they took the cells from your embryo before you were born. In fact, they took the cells even before the embryo began to differentiate into its different organs and tissues. That way the cells kept the maximum capacity for growth and differentiation, and were as free as possible from the damage that began to accumulate as soon as the somatic tissues were formed. The cells for your own future fraitches were taken from your embryo even before they took the germ cells that might eventually be used to make your babies.

The population problem had been foreseen at once. This was when the idea of limited reproduction and Timed Ones was adopted. No one had really objected to the idea that indefinite life span needed to be sacrificed for the opportunity to reproduce. Why, it even mimicked the reason that ageing had evolved in the first place, but for a rather different reason. The evolution of ageing had followed the Darwinian imperative in a world where life was risky and reproduction must be afforded a higher priority than a durable soma. In our own world, Miranda mused, life has become so secure that the luxury of reproduction must be paid for by the giving up of life.

At first, Timed Ones were simply denied further fraitches, condemning them to undergo the customary ageing process. But as ageing became less and less a part of the collective memory, it was decreed that to let someone age was a cruel and unusual punishment, and the first Capsules were used. Once again, some trial and error had been needed. The first Capsules were crude affairs that simply detonated at a precisely timed date to cause sudden death. Living with the knowledge of one's date of death had proved insufferable, as Miranda found had been well known in the barbaric days when certain regions of the world still punished criminals with execution. So the time-window Capsule was developed with the 40–50-year randomiser. A Timed One now held a solemn public farewell at the fortieth anniversary of Capsule implantation and thereafter lived, as Miranda had done, knowing that death could, at any time during the next 10 years, occur within 5 days. The 5-day period had been agreed as providing the best compromise between too sudden and too protracted a dying. The contents of the Capsule had also been modified to ensure a painless but progressive loss of faculties, so that death, when it came on the fifth day, was not too unwelcome an intruder.

'Why 40–50 years?' Miranda had once asked, but she understood the reasons clearly enough now. An early communal decision had been made to entrust the highest executive powers to the Timed Ones, in the knowledge that they alone lacked vested interest in manipulating society to their individual long-term advantage. In some ways, they were like the councils of elders of the ancient times. Forty years was necessary to ensure continuity of government and to give time for the ablest Timed Ones to be elected to, and serve on, the Council. But also, Miranda had noticed with shock when she saw her reflection in a mirror a year ago, it was because 40 years was long enough for the early signs of ageing to reveal themselves. She had been amazed when she caught sight of her first grey hair, and in the weeks that followed, she had run her fingers wonderingly over the beginnings of the first tiny wrinkles in the soft skin at the base of her neck.

'What will Lara make of that, I wonder,' thought Miranda. Lara had always seemed to her a bit of a cold fish. Miranda had secretly felt a small hurt, but no great surprise, that Lara appeared repelled by Miranda's status as a Timed One and that she showed no acknowledgement of the sacrifice that had permitted her own being.

Five years after becoming a Timed One, when her perusal of the Great Archive had told her much that she had wanted to know, Miranda had been invited to join in one of the Special Projects that concerned the future of the race.

The Special Projects were not, of course, the province of the Timed Ones alone. There were too few Timed Ones to make this feasible and the research needed long application of effort. But each Special Project was directed by a team of three Timed Ones, who recruited replacements as the need arose. One of the Special Projects was concerned with the worrying tendency in individuals, after a number of fraitches, for their bodies to grow misaligned. Limbs curved, or were of slightly differing lengths. Organs grew slightly out of proportion. Surgery could repair most of these defects, but a surer measure to correct the drift of the body plan was desired.

Another Special Project considered the problem of mutation in stem cells grown in the test tube. To date, the number of fraitches undergone by the longest-lived individuals had not severely tested the biotechnological capabilities of the cell technicians, but the time would undoubtedly come when the mutation question would have to be solved, especially since natural selection, with its 'purifying' action to eliminate deleterious mutations, was now effectively inoperative. But could a force as fundamental as natural selection really be neutralised?

This was the question that prompted Miranda's eventual choice of Special Project. The world's population size was strictly controlled at a comfortable 100 million, large enough to maintain diversity and farm the arable lands, but small enough to manage without crowding or intensive agriculture. Mating was largely by random individual choice and monitored for genetic compatibility. The former races had merged, but diversity of colour, shape and size continued, favoured by a culture of unprecedented harmony and tolerance. The choosing of the time and partner for making a child was governed by many factors, none of which had so far revealed a discernible pattern. But lately, Miranda had learned, a new trend was appearing. People were choosing to make children earlier. And not only were first children being made earlier, but second ones too. The age at which people were becoming Timed Ones was growing younger. The trend was not yet statistically significant in view of the smallness of the sample, but it was definitely suggestive.

If the trend was genuine, was it caused by a genetic or a psychological change? This was the question that Miranda's Special Project had been addressing.

'Not my Special Project any more,' mused Miranda. 'But I would have liked to know.' It had intrigued her that, if the trend was real, and continued, life spans would grow shorter again. The human race might even evolve to conform to the old fallacious idea that each of us has a reproductive duty to the species and that when this is fulfilled it is time for us to age and die. Miranda chuckled silently to herself at this quaint thought.

'Only a few hours left,' she thought, and slowly opened her eyes. The room, dimmer now, was as it had been before her reverie. Nico was sitting quietly, close beside her couch. Helen, Prato and Cesar, friends of many, many years, were watching from across the room. 'But where is Lara?' Miranda muttered crossly. 'She ought to be here by now.'

A noise of the door opening disturbed the silence and all at once there was Lara, beautiful as ever. Miranda was sad, suddenly, that she could no longer see the red gleam of Lara's long hair, its colour inherited from Gregor, whose tangled locks and beard had always flamed in the light of the sun.

'Forgive me, I am late,' Lara said directly to Miranda, and Miranda could not help but notice the shudder of distaste with which Lara took in Miranda's occasional grey hair and the fine wrinkles that now showed more clearly on Miranda's skin. 'I hope you will not mind, but I have brought Frederic.'

The shock Miranda felt must have revealed itself on her face. To bring a stranger to a dying was unthinkable. For a moment Lara looked painfully awkward, but her face softened into a nervous smile. 'I wanted him to meet you. We are making a child together.'

Miranda closed her eyes with a soft smile and the room was still.

NOTES

Chapter 1 The funeral season

1. The incidence of cardiovascular disease varies from country to
   country, being notably much lower in Japan and other far-eastern
   countries. The causes of these regional variations are not known
   for certain, but are highly likely to involve differences in
   traditional kinds of nutrition. Heart disease rates in the United
   States and Europe are beginning to decline for reasons as yet
   unknown, but which may involve reduced fat intake in the diet.

Chapter 3 What's in a name?

2. There is a curious exception in the case of *Antechinus stuartii*, a
   little Australian rodent that was discovered quite recently to die
   rather suddenly after reproducing. We shall look more closely at
   why this happens and at what it means for ageing in Chapter 11.

Chapter 5 The unnecessary nature of ageing

3. There is evidence from study of the genealogies of European royal
   families that the offspring of older parents may, statistically, have
   slightly shorter lives. The difference is small, however, and may
   arise from minor genetic abnormalities that do not produce any
   obvious physical effect.

Chapter 6 Why ageing occurs

4. Ageing in plants tends to occur differently from ageing in animals
   because in plants the germ-line is often distributed through the
   plant tissues instead of being concentrated in gonads. This is why
   many plant species can be propagated from cuttings without
   obvious limit. Ageing of trees is often the result of mechanical
   constraints associated with physical size. The ageing of annual
   plants is yet another story (see Chapter 11), since these are species
   that indulge in 'big bang' reproduction. The oldest living creatures

are trees, of which the bristlecone pine and Tasmanian holly are the longevity champions, some specimens estimated to be well over 1,000 years old. However, it is unlikely that any of the individual living cells in these specimens is as old as the brain cells in human centenarians, these brain cells having been formed over 100 years ago.

5. I have to say that this question is less likely to be asked by the inhabitants of Navrongo. Abundance of energy has not been the usual state within most of our own species' experience of life on earth, and it is not the natural state for a great many other species either. Many animals regularly die from lack of food, and many people do too. In temperate climates, the need to keep warm in winter is an energy challenge that small animals and old people know only too well.

Chapter 7   Cells in crisis

6. We know now that rapid wound repair often leads to *more* scarring, which is why wounds in old people tend to heal better, albeit more slowly, than wounds in young people.

7. It should also be noted that within each age group there was a considerable amount of variation in the numbers of cell population doublings (CPDs) that each person's cell could achieve. We saw in Chapter 3 that many measures of ageing are intrinsically variable, and cell replicative life span is the same.

8. This gene was discovered at more or less the same time by three different groups and is also widely known as *p21* and *waf-1*.

9. Kindly pointed out to me by Howard (Sid) Thomas, plant biologist at the Institute for Grassland and Ecological Research in Aberystwyth, Wales.

Chapter 8 Molecules and mistakes

10. In fact, Einstein's letter was largely ignored for six months until Rudolf Peierls and Otto Frisch, two more refugees working in Birmingham, England, calculated that only about a kilogram of uranium-235 would be required to form a critical mass. Then began in earnest the race to build the bomb, known as the Manhattan Project.

11. It is of interest to note that the ancestor of the mitochondrion is thought to have been a free-living cell which about a billion years ago found life more comfortable inside another cell and has earned its keep ever since by being the useful converter of energy.

12. Proteins also make up an important fraction of the 'extra-cellular matrix' of the body, which comprises structural materials like collagen.

13. There is a suggestion that faulty chaperones are responsible for prion diseases like Creutzfeld-Jacob disease and the infamous bovine spongiform encephalitis, otherwise known as mad cow disease. Prion diseases are unusual in that the infectious agent appears to be a protein without any genetic material. The conundrum is how a protein can replicate itself in the absence of nucleic acids (DNA or RNA). If the prion is a bad chaperone, and if the bad chaperone can convert good chaperones into bad chaperones like itself, this might resolve the conundrum. It is even possible that bad chaperones have something to do with neurodegenerative conditions like Alzheimer's disease.

Chapter 10 Organs and orchestras

14. A sperm is not required to activate an egg cell and, in fact, the egg is ready prepared with all that is needed to see it through its first few divisions. The genes from the father, carried by the sperm, do not get activated until later. The development of an egg without a sperm is called parthenogenesis and some animals can actually reproduce this way. There is a species of lizard that always reproduces parthenogenetically and has no males. Mostly, however, an activated, unfertilised egg cannot complete the developmental sequence. It is rare for an unfertilised mammalian egg to become activated, but frog eggs can be activated just by a pin prick.

Chapter 11 Menopause and the big bang

15. Some laboratory strains of mice that have been selected over many generations for unnaturally high rates of reproduction also run out of eggs early. Wild mice retain fertility for longer.

16. The geneticist J.B.S. Haldane is reputed to have declared, over a pint of beer in an English pub, that he would willingly sacrifice himself for two siblings, or eight cousins. Evolutionary biologist William Hamilton would later put this idea on a more secure scientific footing when he devised the mathematical theory of kin selection, an elegant concept that explains the otherwise puzzling evolution of certain altruistic and self-sacrificial behaviours.

Chapter 12   **Eat less, live longer?**

17.   Offspring are most likely to be eaten in conditions where their survival is threatened, and when the parent becomes stressed. This is thought to reflect pressure of natural selection not to waste resources, and it is therefore unsurprising that the young get eaten in conditions of food shortage. As well as eating her young after they are born, a pregnant female can also recycle the resources invested in offspring before a litter is born by resorbing her foetuses part of the way through pregnancy. This is more energy efficient than aborting them and it indicates just how important managing the energy budget may be to a small animal.

Chapter 14   **The Genie of the Genome**

18.   You might wonder why women with two copies of the X-chromosome and men with only one do not suffer from a gene dosage problem. The reason is that in women each cell permanently shuts down one of its X-chromosomes. The fact that this neat trick evolved just goes to prove how much gene dosage matters.

BIBLIOGRAPHY

The aims of this bibliography are, firstly, to direct the interested reader to sources of detailed information and, secondly, to acknowledge some of the works which have particularly influenced my own thinking about old age. There is already a huge literature on ageing and my bibliography must, of necessity, be highly selective. The resources of public libraries and booksellers will add to what is listed here, and those with access to the Internet may find it helpful to conduct their search through this medium. Older as well as younger people are increasingly finding the Internet an invaluable (and easy to use) route to knowledge, and it is one that can particularly benefit anyone whose mobility may be restricted or who lives at some distance from bookshops and libraries.

As regards the medical aspects of ageing, there are several excellent, although heavyweight tomes which can be consulted for comprehensive information on the frailties and diseases of old age. The *Oxford Textbook of Geriatric Medicine* (Oxford University Press), edited by J. Grimley Evans and T. Franklin Williams (former director of the U.S. National Institute on Ageing), is one; *Brocklehurst's Textbook of Geriatric Medicine and Gerontology* (W. B. Saunders), edited by Raymond Tallis, Howard Fillit and John Brocklehurst is another. *Epidemiology in Old Age* (British Medical Journal), edited by Shah Ebrahim and Alex Kalache (now chief of ageing at the World Health Organisation in Geneva), is a more manageable volume that gives a broad overview of health and illness in old age.

From the biological perspective, *Biology of Ageing: A Natural History* (Scientific American Library), by Robert Ricklefs and Caleb Finch, gives an accessible account of how ageing affects different species, my one quibble with this otherwise excellent book being that the authors refer to the disposable soma theory with the less familiar name of 'wear and repair' theory. The cellular and chemical factors which are responsible for ageing are described mainly in the current research literature, and the latest information summarised in Chapters 7 and 8 is not easily found in any single source. However, good overviews can be obtained from *Biology of Ageing: Observations and Principles* (Prentice Hall) by Robert Arking, *Longevity, Senescence and the Genome* (Chicago University Press) by Caleb Finch, *Understanding Ageing* (Cambridge University Press) by Robin Holliday, and from a series with the title *Handbook of the Biology of Ageing*, the latest of which was edited by Edward Schneider, John Rowe and Thomas Johnson, and published by Academic Press. A volume entitled *Human Longevity* (Oxford University Press) by David Smith is also a helpful and succinct resource.

The revolution in longevity around the world is characterised most fully in the *United Nations Demographic Year Books*. A small volume entitled *Longevity: to the Limits and Beyond* (Springer Verlag), edited by Jean-Marie Robine and others, presents a number of individual perspectives on the trends in longevity. *Ageing in Developing Countries* (Oxford University Press) by Ken Tout, co-ordinator for special projects with HelpAge International, is a valuable discussion of the demographic challenges facing third world countries. *Too Old for Health Care?* (Johns Hopkins Press), edited by Robert Binstock and Stephen Post, is an informative and thought-provoking collection of papers addressing controversies in medicine, law, economics and ethics that relate to the challenges of a greying society.

Within the realm of literature, there is an enormous amount of material that touches on issues raised in this book. William Shakespeare's love sonnets address the anguish of ageing many times in different ways. There has seldom been as eloquent a study of the emotions associated with ageing as these. *An Old Man's Love* by Anthony Trollope is a remarkable book, well worth reading for its sensitive analysis of intergenerational issues, set in an earlier time. *A Bed by the Window* by M. Scott Peck, bestselling author of *The Road Less Travelled*, is a crime novel, located in a nursing home, that movingly and cleverly succeeds in challenging stereotype views of ageing. *After Many a Summer* by Aldous Huxley is a witty and cynical account of a millionaire's quest for personal immortality, that draws brilliantly on the scientific theories about ageing that were current at the time the novel was written in the 1930s. At the risk of spoiling the final twist – if you have not read it already – Huxley weaves together three strands: Metchnikoff's idea that ageing is due to intestinal microbes, a suggestion published by Bidder that fish like carp which grow indefinitely do not age, and the theory of 'neoteny' (for which, incidentally, there is good evidence) which suggests that during the evolution of humans from apes we stretched the juvenile phase of the ape life span so that, in effect, human beings are like immature chimpanzees.

As regards specific sources, the quotation from John Grimley Evans about the difference between ageing and disease (page 24) is from *Research and the Ageing Population* (Ciba Foundation Symposium 34), a volume which gives a broad, multidisciplinary overview. Richard Peto's assertion that there is no such thing as ageing, and cancer is not associated with it, was published in *Age-Related Factors in Carcinogenesis* (International Agency for Research on Cancer) in 1986. The Baltimore Longitudinal Study of Ageing, discussed on pages 26–28, was described by Nathan Shock and others in *Normal Human Ageing: The Baltimore Longitudinal Study of Ageing* (U.S. Government Printing Office, NIH Publication No. 84–2450). The Edinburgh sea anemones, which died all at the same time after many years of life without any signs of ageing, are described in *The Biology of Senescence* (Churchill Livingstone, 3rd edition) by Alex Comfort (and yes, this is the same Alex Comfort who wrote *The Joy of Sex*). Daniel Martínez experiments with hydra are recorded in the journal *Experimental Gerontology* (volume 33, pages 217–25). The analysis of Jeanne Calment's family records by Jean-Marine Robine and Michel Allard can be found in *Facts and Research in Gerontology 1995* (pages 363–67). The life spans for various animal species recorded in Table 4.2 were drawn chiefly from Comfort's book mentioned above, but corrected and updated with the kind assistance of Steven Austad, whose book *Why We Age* (Wiley) is well worth reading. The history of wrong ideas about why ageing occurs is summarised in a paper by myself and Thomas Cremer in the journal *Human Genetics* (volume 60, pages 101–21); this paper also describes in detail the far-sighted ideas of August Weismann including his prescient prediction, sadly ignored, that somatic cells have finite replicative life spans.

The *Nature* paper first describing the concept of the disposable soma theory was published in volume 270, pages 301–4. Later accounts of this theory, including its more recent development can be found in the *Philosophical Transactions of the Royal Society, Biological Sciences* (volume 332, pages 15–24, and volume 352, pages 1765–72); the latter papers also describe the relationship of the disposable soma theory with Medawar's mutation accumulation (genetic dust-beneath-the-cupboard) theory and Williams' pleiotropic gene theory. Metchnikoff's views on the role of intestinal microbes and the benefits of the Bulgarian bacillus are presented in his 1907 book *The Prolongation of Life* (Heinemann). The evolution of soma in the Volvocales is discussed by Vassiliki Koufopanou in *The American Naturalist* (volume 143, pages 907–31).

The discovery of the Hayflick Limit is described by Hayflick himself in *How and Why We Age* (Ballantine), and the saga of Carrel's infamous chick cell cultures is chronicled by Jan Witkoski in *Medical History* (volume 24, pages 129–42). Our work on the ageing of gut stem cells is to be found in *Experimental Cell Research* (volume 241, pages 316–23). Evidence that lymphocytes in the immune system also show a Hayflick Limit, and this may be linked with the development of AIDS, is reviewed by Rita Effros and Graham Pawelec in *Immunology Today* (volume 18, pages

450–54). The discovery of the gene associated with Werner's syndrome was reported in *Science* (volume 272, pages 258–64).

The basic aspects of organisation and chemistry of cells can be found in any good high school biology textbook, and a particularly clear account of the history of these discoveries is given by John Cairns in his book *Matters of Life and Death* (Princeton University Press). The network model developed by Axel Kowald and me to study interactions between different biochemical causes of cell ageing can be found in *Mutation Research* (volume 315, pages 209–36) and *Experimental Gerontology* (volume 32, pages 395–99). The details of the changes that occur in the ageing body, described in Chapter 9, can mostly be found in textbooks of geriatric medicine, such as those mentioned earlier in this bibliography. Mary Ritter's explanation for the shrinking of the thymus can be found in a paper by Andrew George and her in *Immunology Today* (volume 17, pages 267–72). The biology of cancer is lucidly described in Cairn's book, just mentioned. The connection of telomeres and telomerase with ageing and cancer is reviewed in *Telomeres and Telomerase: Ciba Foundation Symposium 211* (Wiley), edited by Derek Chadwick and Gail Cardew. Work showing that telomerase can apparently push somatic cells to divide through the Hayflick Limit was reported by Andrea Bodnar and others in *Science* (volume 279, pages 349–52).

The biology of the marsupial mice, which indulge in big bang reproductions, is presented in *Evolutionary Ecology of Marsupials* (Cambridge University Press), by A.K. Lee and Andrew Cockburn. The biological basis of menopause is described by Roger Gosden in his books *Biology of Menopause* (Academic Press) and *Cheating Time* (Macmillan) and its evolutionary origins are reviewed in my paper in *Philosophical Transactions of the Royal Society, Biological Sciences* (volume 352, pages 1765–72), where reference to additional sources may be found. Hill and Hurtado's account of the killing of orphans by the Ache people is in their book *Ache Life History* (Aldine de Gruyter). The transition of post-menopausal women to a new, more highly respected status in traditional West

African society is described in *Female and Male in West Africa* (George Allen & Unwin), edited by Christine Oppong. This book also describes very well the broader social context of life in communities such as that around Navrongo.

Calorie restriction and its life-extending effects in mice and rats are clearly summarised in articles by Brian Merry in *Reviews in Clinical Gerontology* (volume 1, pages 203–13, and volume 5, pages 247–58). *The Retardation of Ageing and Disease by Dietary Restriction* (Charles C. Thomas), by Richard Weindruch and Roy Walford, provides more extensive research data on this subject. Adverse effects of foetal undernutrition on health in middle and late age are described by David Barker in *Foetal and Infant Origins of Adult Disease* (British Medical Journal). Gender differences in longevity are examined in the book *Human Longevity* by David Smith, mentioned earlier, which contains information of the differential attention paid to male and female children in the Indian sub-continent. The explanation in terms of the disposable soma theory of why women live longer than men, as presented in Chapter 13, is original to the present book. The evidence that virgin male and female fruitflies live longer than mated flies in spite, in some experiments, of having their external genitalia ablated by cauterisation with a hot wire (!), may be found in the *Journal of Insect Physiology* (volume 43, pages 501–12) and *Nature* (volume 338, pages, 760–61, and volume 373, pages 241–44).

Evidence for inheritance of human longevity is in a report by Matt McGue and others in the *Journal of Gerontology, Biological Sciences* (volume 48, pages B237–B244). Strategies for investigating genetic factors affecting human longevity were described by François Schächter, Daniel Cohen and me in *Human Genetics* (volume 91, pages 519–26). The study by Schächter and others of gene polymorphisms in French centenarians was reported in *Nature Genetics* (volume 6, pages 29–32). A review by Caleb Finch and Rudolph Tanzi of genetic factors affecting longevity and age-related diseases, such as Alzheimer's disease, can be found in *Science* (volume 278, pages 407–11).

Sources of information on potential forms of 'Wonka-Vite' are widespread but the reliability of the articles is highly variable, as discussed in Chapter 15. The search for drugs that combat Alzheimer's disease is described by Jean Marx in *Science* (volume 273, pages 50–53). Joan Smith-Sonneborn's experiments showing how the life span of *Paramecium* was extended by inducing DNA repair with ultraviolet light was reported in *Science* (volume 203, pages 1115–17), and Gordon Lithgow's studies showing that heat can make nematode worms live longer were described in *Proceedings of the National Academy of Sciences USA* (volume 92, pages 7540–44).

The poem by Seamus Heaney on the dedication page is from *Seeing Things*, that by Edna St Vincent Millay on page 184 is from *Collected Poems*, and the extracts from poems by U.A. Fanthorpe (page 196), Sheenagh Pugh (page 212) and Ira Gershwin (page 39) are from *Safe as Houses*, *Selected Poems* and *Porgy and Pess*, respectively, by kind permission of the copyright holders.

INDEX